Also by Kenneth S. Deffeyes

Hubbert's Peak: The Impending World Oil Shortage

Beyond Oil

Beyond Oil

The View from Hubbert's Peak

KENNETH S. DEFFEYES

HILL AND WANG

A division of Farrar, Straus and Giroux

New York

Hill and Wang
A division of Farrar, Straus and Giroux
19 Union Square West, New York 10003

Distributed in Canada by Douglas & McIntyre Ltd.
Printed in the United States of America
First edition, 2005

Library of Congress Control Number: 2004116475
ISBN-13: 978-0-8090-2956-3
ISBN-10: 0-8090-2956-1

Designed by Patrice Sheridan

www.fsgbooks.com

1   3   5   7   9   10   8   6   4   2

To Michael
For what he taught me

Anyone who believes that exponential growth can go on forever in a finite world is either a madman or an economist.

—KENNETH E. BOULDING

Contents

Preface

We are facing an unprecedented problem. World oil production has stopped growing; declines in production are about to begin. For the first time since the Industrial Revolution, the geological supply of an essential resource will not meet the demand.

There has been plenty of warning. In 1969, M. King Hubbert, an American geologist, published predictions of future world oil production.[1] Hubbert predicted that annual oil production would follow a bell-shaped curve; the curve became known as "Hubbert's peak." The more optimistic of his two estimates in 1969 placed the world's total oil endowment at 2.1 trillion barrels and peak production in the year 2000. My best current estimate (detailed at the end of Chapter 3) puts the total oil at 2.013 trillion barrels, peaking in 2005.[2] Whichever of us is correct, or even if we are both wrong, we are not very wrong. Wherever the peak, the view is not good.

My own interest in the oil supply problem began in 1958, when I started work at the Shell research lab in Houston. Hubbert enjoyed superstar status at the Shell lab, for many reasons in addition to his oil predictions. I enjoyed working at the lab, and I enjoyed getting to

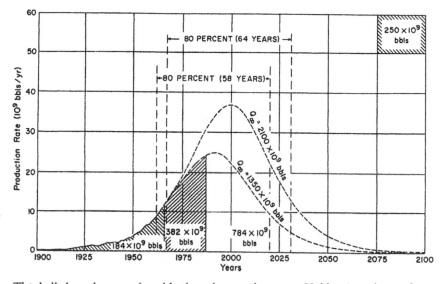

This bell-shaped curve of world oil production, known as Hubbert's peak, was first published by M. King Hubbert in 1969. The more optimistic of the two alternate versions of the curve predicted that production would peak around the year 2000. Hubbert made this prediction using an early, primitive version of his method that required educated guesses for the amount of total available oil. As the history of exploration and production progressed, Hubbert was no longer dependent on educated guesses. To this day, however, critics sometimes state incorrectly that all Hubbert-type estimates require previous independent estimates of the total available oil. (Hubbert, "Energy Resources," p. 196)

know Hubbert. By 1963 it was clear that the oil business (in Houston, it's the "awl bidness") would change enormously by the year 2000, when I was supposed to retire. My own analysis of Hubbert's numbers caused me to leave the petroleum industry prematurely.

Emotionally, it was not easy to leave. I grew up in the oil patch; my father was a first-generation petroleum engineer. I was born in Oklahoma, am part Chickasaw, and can remember the Dust Bowl. When I was about ten years old, I decided to become a petroleum geologist. While I was in high school, a nice piece of help came along: Two geologists, Jack Fanshawe and Paul Walton, took the time to help me learn mineralogy. My undergraduate education (Colorado School of Mines) and graduate education (Princeton) were focused on methods useful in oil exploration and production. I held a sequence of summer jobs that taught me the dirty-fingernail side of the oil industry. For a while, I had a bumper sticker that said, "Oilfield trash, and proud of it."

After I left Shell, I taught for a couple of years each at Minnesota and Oregon State, and then joined the faculty at Princeton. Teaching and research were rewarding, and I could continue consulting part time on oil and mining. My course titled "Sedimentology" was a camouflaged course in petroleum geology.

This book is about the fuels that come from the earth. A lot of questions about public policy naturally follow. However, my expertise ends where the geology stops. Any opinions I have about the wisdom of increased gasoline taxes or a U.S. oil import tax have the same standing as the opinions of J. Random Citizen. In this book, I try to explain the advantages and constraints on the various fuels from the earth. Deciding on policy is a task for all of us as citizens.

Chapters 1 and 2 are brief statements of the oil supply problem. Chapter 3 is the result of a happy accident: I found an alternative to Hubbert's complicated mathematics. Hubbert's analysis requires pages of differential equations to go from A to B. Going from B to A reaches the same results but requires only three lines of high-school algebra. Chapters 4 through 8 cover natural gas, coal, tar sand, oil shale, and uranium. Chapter 9 is an oddity: Hydrogen is not a fuel that comes from the earth, but there is so much misinformation floating around about hydrogen that I felt compelled to explain further about it. Chapter 10 is an essay about seeing the world through a geologist's eyes; it is what William Safire calls a "thumbsucker."

There was a tremendous flap during 2004 over Shell's downgrading of oil reserves. Those of us who used to work for Shell were particularly surprised; typically Shell was overly cautious about almost everything. I have no private sources of information from inside the major oil companies. Mostly, I try to evaluate what they do, not what they say. For instance, an editorial in the June 21, 2004, issue of *BusinessWeek* complained that the 30 percent increase in oil prices induced only a tiny increase in company exploration budgets. Similarly, U.S. refineries are running close to capacity, but no new refineries have been built since 1976. Oil tanker ships are fully booked, but outdated tankers are being retired faster than new ones are being built. Instead, the industry seems to be hoarding cash, buying back stock, and paying out dividends. What is going on? Why don't higher prices and increasing demand encourage investment? Suppose, for a moment, that the premise of this book is correct: We have already found most of the oil. Drilling for the few leftovers yields neither fun nor profit. Should the major oil companies drill a string of dry holes just to keep the editors of *Business-Week* happy? If, as I claim, world oil production is about to decline, then there is no point in adding refineries or increasing the size of the tanker fleet.

The major oil companies are not saying publicly that the oil game is over. If there were attractive prospects available, companies would be clawing their way over one another to get the drilling rights. There is important exploration to be done in a number of countries. Even if an oil company signs a contract with either the national oil company or the government itself, in many countries the contracts are unenforceable. Drill a dry hole and nothing happens. Hit a major discovery and suddenly the contract is up for renegotiation.

During the last four years, there has been considerable interest in the financial community about the oil supply problem. Elsewhere, a few conservation-minded people have paid attention. How can you reduce carbon dioxide in the atmosphere? Run out of oil. Politicians and the general public have paid little attention. Here is part of the reason: Some professional petroleum observers state that world oil production will continue to increase until the year 2030. Any publication that pretends to be "fair" feels compelled to present both sides of the story. When the professionals disagree, does that mean there is no real knowledge available? Is it safe to ignore the problem until the professionals agree?

Here's my reply: Doing nothing today is simply betting that Hubbert is wrong. Hubbert's 1956 prediction that U.S. oil production would peak in the early 1970s was essentially correct. Hubbert's 1969 prediction that world oil production would peak around the year 2000 is coming true right now (details are in Chapter 3). Fifteen years ago, we should have started investing heavily in alternative energy strategies. That opportunity is now lost. There is no time left for scholarly research. There is no time left for engineers to develop new machinery.

We have to face the next five years with the equipment designs that are already in production. It's not going to be easy.

Acknowledgments

For detailed comments on the first draft, I want to thank Bill Bonini, Larry Cathles, Robert Deffeyes, Suzy Deffeyes, Sarah Domingo, Immanuel Lichtenstein, Peter Lu, Eldridge Moores, Jason Phipps Morgan, and W. Jason Morgan. Joe Wisnovsky and Thomas LeBien gave close readings to later drafts. However, I am responsible for any errors or omissions. (I didn't always take their advice.)

Disclaimer

Much of the material in this book is relevant to the future course of national and world economies, but the book is not intended as an investment guide. I have no expertise in "the dismal science" of economics; my only training in economics was a single undergraduate course fifty-two years ago. To put it more bluntly, my track record with my own investments is best described as spotty.

Beyond Oil

One

Why Look Beyond Oil?

The supply of oil in the ground is not infinite. Someday, annual world crude oil production has to reach a peak and start to decline. It is my opinion that the peak will occur in late 2005 or in the first few months of 2006. I nominate Thanksgiving Day, November 24, 2005, as World Oil Peak Day. There is a reason for selecting Thanksgiving. We can pause and give thanks for the years from 1901 to 2005 when abundant oil and natural gas fueled enormous changes in our society. At the same time, we have to face up to reality: World oil production is going to decline, slowly at first and then more rapidly.

My late-2005 prediction is a statement about the smoothed average of annual oil production. As we go over the top of the smoothed average, individual years, like wine vintages, have their ups and downs. At the time of this writing (in the middle of 2004), the year 2003 had the largest oil production.[1] However, production in 2003 was only 3 percent larger than the production in 1998. That's not 3 percent *per year*; it is only 0.6 percent growth per year. World oil production has now ceased to grow. Decline is the next step. The picnic's over.

My prediction of the world oil production peak is based on the methods that M. King Hubbert used in 1956 to predict the 1970 peak of U.S. oil production.[2] (Hubbert's method is explained in detail in Chapter 3). Even before I published my 2001 book *Hubbert's Peak: The Impending World Oil Shortage*, half a dozen petroleum geologists had already published similar conclusions.[3] Hubbert published his own world oil prediction in 1969; the more optimistic of his two scenarios placed the peak in the year 2000. On the other side of the debate, the U.S. Geological Survey published enormously more optimistic estimates for U.S. and world oil potential.[4] Other analysts, using the USGS results, have predicted that world oil production will not peak until 2036.[5]

Thanksgiving

Between 1901 and 2005, oil and natural gas transformed our society. In 1901, the Spindletop well, near Beaumont, Texas, changed the basis of

M. King Hubbert was well known among geologists for several innovative ideas, usually backed by mathematical analysis. After his prediction of the U.S. oil production peak came true, conservationists considered Hubbert to be a folk hero. (© George Tames, *The New York Times*)

the petroleum industry. It opened the enormously productive area of the Gulf Coast, pushed U.S. oil production to first place in the world, displacing Russia, and introduced the rotary drilling rig, which is still the standard today. Moreover, the flow of water, oil, or gas into the well bore could be controlled. The last point, control of subsurface flow, removed the "gusher" scene from the routine oil business, except for accidental failures of the system.

My parents, born in 1903 and 1904, grew up on farms in two different parts of Oklahoma. My mother claimed that the automobile was the major element of change, introducing social as well as personal mobility. Before the automobile, there were weekly trips in a horse-drawn wagon to the nearest small town to sell farm produce and buy supplies. The Ford Model T enabled my parents to find their way to Oklahoma A&M, now renamed Oklahoma State University. One of my aunts claimed that the major motivation for higher education was to get off of those farms.

The automobile and the oil business were made for each other. From 1859 through 1908, the major petroleum product was kerosene for lanterns. After that, automobiles and trucks became a rapidly expanding market. Oil refineries were gradually improved to turn a bigger fraction of the crude oil into gasoline. Oil exploration and production was a worldwide enterprise.

The year 1903 saw the Wright brothers' first flight. The high energy content per unit weight of gasoline was even more important for airplanes than for cars. High-grade aviation gasoline became a profitable product, even though aviation was a smaller market than automobiles. With the 1945 introduction of the jet aircraft engine, a kerosenelike product known as JP-4 became important. Mobility is now global. It is possible today to travel door-to-door from a street address in any city in the world to any other city in less than twenty-four hours. It's not just people: Mail, computer components, mangoes, and asparagus move by air freight.

Meanwhile, back on the farm, agriculture was transformed. Tractors replaced horses, produce was trucked to wider markets, and mineral fertilizers became widely available. The Green Revolution of the 1960s used improved seeds, pesticides, and mineral fertilizers to make famine obsolete. It wasn't "organic," but it sure beat death from starvation. A measure of the importance of oil and gas: 80 percent of an Iowa corn farmer's costs is, directly and indirectly, the cost of fuels.

A whole industry emerged for making products derived from oil: petrochemicals. An astonishing range of plastics, fibers, solvents, pesticides, and coatings are made from oil and natural gas. My guess is that petrochemicals will be the last, best use of petroleum as it becomes scarce. Using oil as premade building blocks for organic chemistry is better than burning it for fuel. When I am offered an unnecessary plastic bag at the grocery store, I reply, "No, save an oil well."

I visited England in 1967 and again in 1985, and I was amazed at the changes that had occurred in the meantime. The legendary poorly heated housing became warmed by gas newly discovered in the North Sea. A feeling close to hopelessness had been replaced by a whiff of prosperity. The wealth from oil and gas is very irregularly distributed throughout the world, both among countries and among individuals. That's partly geology and partly politics. Big oilfields occur wherever all of the several required geological conditions are "just right." Our subdivision of the world into ever-smaller sovereign countries and the success of private-property systems concentrated much of the oil wealth into a few hands. An example: During my stay at the University of Cambridge in 1985, I needed a book in their main library. As I entered the building, I had the weird feeling that I wasn't in England anymore. The Cambridge main library looked American. I found a hallway display that explained it. The library had been built during the Great Depression, with Rockefeller money. Oil money.

I'm not claiming that oil and gas generated all the goodies of the twentieth century. Computers and telecommunications succeeded in part because they used diminishing amounts of energy. Silicon is the second most abundant element in the earth's crust, so the supply of materials was never a problem. Well, there was some help from the oil industry. The patent rights for the integrated-circuit chip were shared by Texas Instruments and by Intel. At the time of the invention, Texas Instruments was a service company, conducting seismic surveys in the search for petroleum.[6]

Because I am a geologist, I have my own narrow reasons to give thanks for the growth of the oil industry. Well-staffed and well-funded research laboratories were developed by the major oil companies. To a geologist, the miracle was having top-grade physicists, chemists, and mathematicians on the staff, willing and eager to discuss geological problems. Two technological innovations were important because both of them generated objective, noninterpretive, data:

- Sensors were sent down wells, at the ends of electrical cables, to measure rock properties. Known as "wireline logs," they were run on virtually every well and are a rich source of data.
- Sound waves generated at the surface and reflected back from subsurface rock layers produced cross sections similar to the sonograms used in physicians' offices.

In addition to the research results generated internally by the oil companies, important work was done by national and state geological surveys. Because geologists were being hired by oil companies, geology departments in colleges and universities thrived. In addition to the major universities, excellent geology departments developed at colleges like Macalester, Bryn Mawr, Smith, Beloit, and Williams. The overall effort moved petroleum geology into first place intellectually among the earth sciences, with the most complete understanding of any geological resource.

What Happens Next?

So the big news is: World oil production has ceased growing, and by the year 2019 production will be down to 90 percent of the peak level. This is not your standard news story. Reporters, editors, and most readers expect that any news story goes away within days, occasionally months. The peak-oil story not only stays around; each time that we adapt to a lower level of production, the production falls again. We may have fallen short on oil, but we are doing great with lead and zinc, which are abundant and cheap—if you happen to want them.

On a fifteen-year time scale, I have no doubt that human ingenuity will find adequate energy sources with nice adjectives like "renewable," "nonpolluting," "sustainable," "alternative," "organic," and "natural." For the five-year time scale, we have a shortage of good adjectives.

"Diesel," "coal," "nuclear" don't sound warm and fuzzy. There has been plenty of warning. Some predictions were issued more than thirty years ago. Why wasn't there a strong effort toward developing a new energy economy? My own feeling is that editors and news directors thought that their audiences weren't interested in yet another Chicken Little story. No politician was going to run on a platform promising "blood, sweat, and tears."

Am I now promising war, famine, pestilence, and death? If we can keep the petrochemicals industry healthy, we might avoid the pestilence part. The other three are serious possibilities. While a new energy economy is being implemented, there will have to be some sort of regulation of scarcity. Virtually all economists visualize it as price increases that bring supply and demand into a new equilibrium. That outlook is widespread; it must be something that Gerber puts in baby food. Historically, President Nixon regulated the oil price. President Roosevelt had us carrying little red and blue gasoline ration coupons. When the situation gets serious, there will be immense political pressure to "do something."

What can we expect on the five-year time scale?

- In winter, carrots, potatoes, cabbage, turnips, onions, and beets: local produce that does not have to be flown up from the southern hemisphere. Get acquainted with parsnips and rutabaga.[7]
- A premium on getting out of a long daily automobile commute: relocating jobs and housing, mass transportation, teleconferencing.
- Trade in your Hummer or Porsche Cayenne; find some other way of publicizing your testosterone.
- Agriculture in the Third World will suffer greatly with too little diesel fuel and mineral fertilizers; starvation is on the agenda.
- A scramble for high-efficiency diesel engines, for wind farms, and for any undeveloped hydroelectric sites.
- Agony over opening new nuclear or coal plants for generating electric power.
- Concerns over ethanol and hydrogen as net losers: technologies that consume more energy than they produce.

Within the oil industry, technological leadership has largely passed from the major oil companies to service companies like Schlumberger and Halliburton.[8] The major oil companies are coming to resemble large service companies with attached merchant banks. Ownership of the oil production is a diminishing component of their income.

Again, in my narrow view as a geologist, I have a concern about the talent pool for finding the remainder of the oil. Colleges and universities responded to student interests and to availability of research funds by putting the word "environmental" on many of their courses. Environment has become a higher calling, much less grubby than completing an oil well. About six schools in North America currently offer training adequate for obtaining an entry-level job in the oil industry.[9] Further, senior geologists in the industry have traditionally helped new employees learn the profession. In one sense, the oil industry is relatively young. My father was among the first generation of petroleum engineers; I'm only a second-generation oilman. A break in the continuity equates to a tremendous loss of practical knowledge. Matthew Simmons, an energy banker in Houston, calls it "no freshman class."[10]

And what about war, the first horseman of the apocalypse? At a petroleum-supply meeting in Europe during the spring of 2003, I was startled to learn that virtually all the Europeans believed that the Iraq war was simply for the control of oil. At the time, I was evaluating several different reasons for the invasion of Iraq (all of them bad, but I'm a registered Democrat). The major oilfields of the Middle East are located in several countries, but the entire producing area is only a fraction of the size of the United States. If the world oil shortage becomes sufficiently painful, there could be a temptation for a military seizure of the oilfields and the establishment of a "world protectorate." Some bureaucrat, even now, might be gathering euphemisms to justify an ugly scene.

My bottom-line conclusion says that the biggest missing ingredient is political leadership. Leadership does not necessarily imply a wise philosopher-king. Here are two examples. During World War II, Winston Churchill might have been a major daily consumer of brandy. He may not have been facing reality. Ignoring reality was precisely what England needed. I was in Rice Stadium in 1962 when John Kennedy proposed the manned lunar missions. It seemed like a tremendously

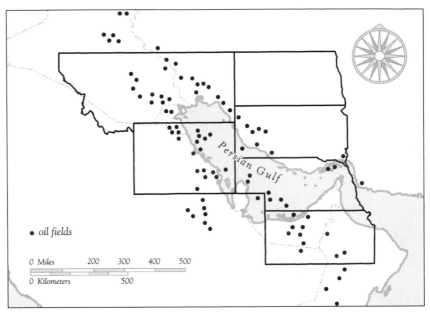

Almost all of the oilfields in the Middle East would fit into the north-central region of the United States. More than a quarter of the world's oil production comes from this limited area. (Jeffrey L. Ward)

good idea at the time. In retrospect, unmanned exploration might have been better. Also, rumor has it that Kennedy was on methamphetamine.

I'm not saying that a chemically boosted head of state is desirable. What we desperately need is a leader who can put words together and convince us that we better get moving.

In short, a small but growing number of petroleum geologists agree that world oil production will reach its peak sometime during this decade. There are plenty of geologists, economists, social scientists, and statisticians who disagree, and some who vehemently disagree. Unkind words have been exchanged. In polite society, I prefer calling them "cornucopians." They feel that human cleverness or the invisible hand of the marketplace will push the peak out twenty or more years in the future. My colleagues and I, indirectly or directly, are followers of M. King Hubbert, who predicted the U.S. oil production peak fourteen years before it happened. More about Hubbert later; for now, the people on my side are "Hubbertians."

Alan Greenspan, chairman of the Federal Reserve Board, has been warning us of a potential natural gas shortage in North America.[11] It won't be fun. There are some options on natural gas: liquefied gas imports, Arctic pipelines, moving some chemical operations—and the associated jobs—overseas, replacing natural gas electric generating plants with coal or (gasp) nuclear plants, and especially conservation. Oil is a bigger and uglier problem than natural gas.

Almost all authors, in their introductions, try to convince readers that their books are important. Despite the risk of traveling that well-worn path, let's try this multiple-choice question:

Who gets credit for causing the collapse of the Soviet Union?

    a.   Ronald Reagan, for promoting Star Wars
    b.   the pope, for being Polish
    c.   Mikhail Gorbachev, for allowing dissention
    d.   the KGB, for abusing the people
    e.   Saudi Aramco, for lowering oil prices

Stephen Kotin points out that the Soviet Union, up to 1985, was exporting two million barrels of oil per day.[12] The hard currency from oil allowed the Soviets to import items that were internally in short supply, from electronics to soap. At that time, Soviet oil production was larger than Saudi production by a factor of three, but Saudi Aramco had much lower production costs. Saudi Aramco resorted to a familiar tactic: a price war. They flooded the world with oil and drove the world price of crude oil below the Soviet cost of production and transportation. The severe shortages of everything that developed within the Soviet bloc are illustrated by this story:

A Polish businessman is going on a trip to Moscow and his wife asks him to bring back notes about the shortages in Russia. He goes into a butcher shop, and there are only a few little scraps of salt pork, so he writes in his notebook: NO MEAT. He then goes into a greengrocer and writes NO VEGETABLES. A shoe store: NO SHOES. After several more shops, a man stops him on the street and asks, "Are you spying for the CIA?" The businessman explains that his wife asked him to take notes. "Don't you know that ten years ago you would have been shot for doing this?" He writes: NO BULLETS.

After six years without hard currency, the Soviet Union collapsed. Control of the world's dominant energy source carries enormous power.

Oil is already a global market. Supertankers move oil to the far side of the world for two dollars per barrel. Our transportation system is almost totally driven by products from oil. As we learned in the late 1970s, an oil shortage can ripple through the economy, lowering our standard of living. My concern is not about our long-term adaptation to a world beyond oil. Through our inattention, we have wasted the years that we might have used to prepare for lessened oil supplies. The next ten years are critical. It's going to be on-the-job training. Learn while doing: not always the most orderly way of adapting.

As noted above, this book is about the fuels, in addition to oil, that come from the earth. I'm a geologist; I don't have any wisdom to share about solar cells, biofuels, or tidal energy. As a preface to the other fuels, the next chapter explains why crude oil production is likely to diminish. If oil were abundant and cheap, there would be no need to develop the other fuels.

Two

Where Oil Came From

When I was young, Sinclair filling stations displayed a dinosaur as a trademark. Like most kids, I thought that oil in the ground might have been squeezed out of the bodies of dead dinosaurs. Later, I wondered whether most oil was formed during the age of the dinosaurs. The real story has several components.

- Basins at the edges of the world ocean—like the Red Sea, the Yellow Sea, and the Black Sea—sometimes develop a circulation pattern that traps marine nutrients. Dead plants and animals and fecal pellets fall out of the surface water, taking nutrients with them. They are normally oxidized by bacteria on the sea floor, which releases the nutrients back into the water. This constant organic fallout from the surface continually hauls nutrients into the deep water. When an arm of the sea has more surface evaporation (from sun and wind) than freshwater input from rain and rivers, a circulation pattern develops that brings in shallow ocean water and exports the nutrient-rich deep water back to the main ocean. If river runoff plus

The Sinclair trademark led many children of my generation to believe that oil came from dinosaurs. This is a restored historic pump, with gas at 16¢ per gallon. (Tom Worner, Associated Press/*Tyler Morning Telegraph*)

rainfall exceeds surface evaporation, then nutrient-depleted surface water is exported back to the open ocean and the deep water becomes a nutrient trap. Wherever the nutrient-laden deep water is mixed toward the surface, surface marine life blooms. Phosphates and nitrates are usually the critical nutrients. The Black Sea comes close to being a modern example. About a thousand years ago, the Black Sea bottom water was fully depleted of oxygen. Sediments with more than 10 percent organic matter were preserved on the bottom of the Black Sea. Organic-rich sediments are not common; less than 1 percent of all sedimentary rocks contain more than 5 percent organic carbon. If organic-rich sedimentary rocks were common, we would be swimming in oil.

- Burying organic-rich sediments in rocks deeper than 7,500 feet takes them into a temperature range (about 175°F) that causes large organic molecules to break into smaller pieces. Molecules with five to twenty carbon atoms are liquids: crude oil. Molecules with fewer than five carbon atoms are gases at room temperature and pressure: natural gas. A depth of 7,500 feet is called the "top of the oil window."

- Burying the sediments, or the oil, deeper than 15,000 feet continues the molecular breaking until the remaining product has only one carbon atom per molecule. That gas, almost pure methane (CH_4), is often referred to as "dry" natural gas. The limit of 15,000 feet is the bottom of the oil window. If you are looking for oil, you need organic-rich sediments that have been buried, at some time in their history, into but not deeper than the oil window.

- After liquid crude oil is liberated within the oil window, it tends to migrate upward. The oil is not soluble in water and it floats upward because it is lighter than water (think salad dressing). Most oil, at least 90 percent, finds its way to the surface as oil seeps. The remaining 10 percent (or less) gets trapped underground in places that are not well connected to the surface, such as domes, fossil reefs, sand lenses, and even a few meteorite impact sites.

- There are no big open caverns at oilwell depths. The oil is trapped in rocks with significant porosity, known as reservoir rocks. ("Significant" is more than 5 percent.) Roughly half of the world's oil comes from sandstone reservoirs. Almost all of the other half comes from limestone and dolomite. Limestone— which is calcium carbonate, $CaCO_3$—is formed mostly by marine plants and animals in ocean surface waters. Dolomite, a calcium-magnesium carbonate, $CaMg(CO_3)_2$, is still not fully understood, but it makes some of the best reservoir rocks.

- The pore spaces in the rock have to be connected to one another so that oil or gas can flow through the rock; the pore connectivity is called permeability. Gas bubbles in a solidified lava flow are examples of pores that are not connected to one another. Permeability depends on the square of the grain size. Doubling the diameter of the sand grains in a sandstone increases the permeability by a factor of four. Increasing the grain size by a factor of ten makes the permeability a million times larger.

- At least one rock layer between the oil reservoir and the surface has to make a leak-tight seal, called a cap rock. The first layer above a commercial reservoir rock is often mistakenly identified as the cap rock, although it usually is simply an oil-saturated but noncommercial reservoir rock. One of my favorite homework problems for my students asked how long it would take for a single leak of one drop per second to empty a billion-barrel oilfield. The answer: 100 million years. The middle of the Cretaceous period was 100 million years ago; many oilfields are older than that. Leak-proof cap rocks are special: very fine-grained mudstones and evaporite layers like halite ($NaCl$, table salt) and anhydrite ($CaSO_4$).

So we have seven essential ingredients in the list. What if one or more of them is missing? You find zero oil. Nature has a funny way of grading exams. You answer seven questions; your grade on the overall exam is the lowest grade that you got on any one of the seven.

Because of the peculiar grading system, very few places on Earth get overall high grades. Everything has to be right, or nearly right. The

Middle East is at the head of the class: excellent source rocks buried into the middle of the oil window, rocks deformed into big domes, good porosity and permeability, and anhydrite cap rocks. Runners-up are West Texas, the North Sea, and a dozen other places that did reasonably well on all seven questions. Most areas flunked—took a zero on one or more questions. Oilfields cover less than 0.1 percent of the continents and continental shelves.[1] A few places are incredibly oil-rich, most areas have nothing. It seems unfair, but that's nature's grading system.

The United States and Russia have long and complex oil production histories. Some chronologies place the discovery of the Baku, Azerbaijan, oilfield before the Drake well of 1859.[2] (In the remainder of this book, I will use the Drake well as a metaphor for the startup of the oil industry. I will not stop each time to explain the pre-Drake alternatives.) Saudi Arabia is the country with the largest annual pro-

This undated photograph shows early derricks near Baku, on the Caspian Sea. Some accounts place the discovery of this oilfield ten years before the Drake well of 1859. (SOVFOTO)

duction, but Russia and the United States are in second and third place. The contrast shows up in the number of producing wells:

	Production (million barrels/day)	Producing Wells
Saudi Arabia	7.7	1,560
Russia	7.4	41,192
United States	5.8	521,070

During the Communist and immediate post-Communist era, the Russian oil industry suffered from deferred maintenance and mismanagement. After the year 2000, high international oil prices and improved maintenance increased Russian oil production. So far, there have been no reports of major new Russian oil discoveries. It remains to be seen whether the renaissance of Russian production is a one-time repair of past problems or will become a growth industry.

Oil Exploration

"Is it oil country?" That's the first question asked about a region. If there are already producing wells in the area, the answer is obviously yes. If the region is either unexplored or if a few dry holes have been drilled, the reply is maybe. Here are ways of getting an answer without drilling a huge number of dry holes:

- Until about 1930, natural oil seeps were an important guide. If you see an iridescent surface film on stagnant water, disturb it with a twig. If it breaks into rigid rafts, that's an iron oxide film. If it deforms smoothly, it's oil.
- Existing dry holes may have encountered noncommercial "shows" of oil. The dry holes could have flunked any of the last four of the seven questions (oil migration, reservoir rocks, permeability, leak-proof seal).

- Major companies sometimes chip in together to drill a "strat test" in a location away from a dome or any other likely oil trap. They split up the samples and each company does its own examination for source rocks that have been in the oil window.

Exploration geologists know that a few dry holes do not condemn a region. If there is reasonable evidence that an area could be oil country, they will keep plugging away until someone turns off the money supply.

Historically, oil exploration was conducted by looking for nearly round domes or elongated anticlines that could trap oil. In arid climates, the surface geological structure is readily visible. Even in the

A tour guide pulls fresh tar from beneath a rainwater puddle at Pitch Lake on the island of Trinidad. (Shirley Bahadur, Associated Press)

jungle, there are usually outcrops a few inches high alongside the streams. In 1953, I participated in the very last round of surface exploration in Wyoming. In Iraq, and possibly in Iran, all of the drilling has been based on surface exploration.

It takes only a few tens of feet of thick tropical soil, or glacial gravel, or young sediments to make surface exploration impossible. The workhorse today is called reflection seismic exploration. The earliest seismic reflections were recorded in Oklahoma in 1922.[3] Near-surface dynamite explosions generated sound waves; sensitive surface-motion detectors picked up the reflected sound. In mountainous areas or in jungles, explosives are still used, although dynamite has been replaced in part by a mixture of ammonium nitrate and nitromethane. I used to be afraid that this recipe would become known outside the industry. Then the late Timothy McVeigh used drums of ammonium nitrate–nitromethane in his Oklahoma City bomb.[4]

On land, explosives have largely been replaced by heavy truck-mounted vibrators to generate the sound waves. The early photographic records were replaced by digital recordings. The calculations

Five Vibroseis trucks simultaneously send a sound signal into the ground without damaging the paved road.

that I did by hand in 1952 were replaced by banks of "big iron" IBM mainframe computers. It wasn't just a matter of replacing me with cheaper computation; seismic processing achieved a complexity that would have been impossible by hand calculation. Despite the improvements, seismic exploration on land is a factor of ten more expensive per square mile than marine seismic exploration.

At sea, the strategy is the same but the equipment is different. Explosives have been replaced by air guns that release high-pressure compressed air through quick-acting valves to form an underwater bubble. Strings of sound detectors are towed on long parallel streamers behind the ship. Everything moves along, twenty-four hours a day, at two or three miles per hour. For a while, even the computer processing was done aboard the ship. Around 1998, the third-largest computer cluster in the world was on a seismic-prospecting ship. Eventually, it became more economical to send the data to computer centers on land,[5] but

A pad beneath the center of each truck transmits the seismic vibration into the ground, in this instance in a grass pasture.

not because it was inconvenient to put the computer on the ship. The human experts could manage more than one ship from a land-based computer station.

From its origins in 1922 up to 1995, all reflection seismic prospecting was done along "lines." A linear string of seismic detectors and the sound source moved along the line. Gradually, lines are being replaced by a rectangular grid of detectors and the sound sources occupy multiple locations within the grid, known as 3-D seismic. Also, use is being made of shear waves as well as compressional waves. Compressional waves are essentially the same as sound waves in air. Shear waves move from side to side and are transmitted only through solids. Liquids and gases have no shear strength and cannot transmit shear waves.

Early seismic prospecting was done solely to detect geological structure: anticlines, salt domes, faults. Even before 3-D seismic became available, seismic lines were used to detect subsurface fossil reefs. With 3-D, the level of detail sometimes is amazing. Small features, such as sand-filled river channels, are imaged.

In mature areas there can be hundreds to thousands of wells in a ten-mile-by-ten-mile square. Virtually all wells are logged by lowering a sensor on a steel cable into the well. A sensor includes detectors for about twenty different physical properties. (In 1927 in France, the Schlumberger brothers first applied the technique. Schlumberger has since grown into a comprehensive oilfield service company.) Most logs in the United States are released to public archives so that even an independent geologist can construct subsurface maps.

Drilling Rights

Ownership of subsurface deposits varies from country to country. When the king owned all mineral rights, miners paid a fee called a royalty to the crown. In most countries today, the government retains title to subsurface metals and petroleum. Oil companies may be granted rights to explore, produce, and sell oil in exchange for a royalty, a percentage of the income from sales. The government percentage has become larger with the addition of excise taxes, export taxes, and joint participation by a (noncontributing) national oil company. Even where exploration and production rights were previously in place, oil operations have been nationalized. The anniversary of the 1938 nationalization of the oil industry is still observed in Mexico. Since then Venezuela, Iraq, Iran, and Saudi Arabia have turned their oil production over to nationally owned oil companies. I don't view this as a moral issue. Governments simply recognized that the oil business was highly profitable and horned their way into our racket.

The United States has a mixed bag of oil and gas rights. Even in the oil patch, the subsurface rights are called "mineral" rights. Most, but not all, of the private land acquired by homesteading or by other means carried both surface and subsurface rights. After the land was acquired, subsurface rights could be bought and sold independently. I inherited from my mother a miscellaneous collection of small shares of mineral rights in south-central Oklahoma. Oklahoma has a "forced pooling" law that requires a driller to obtain at least 63 percent of the mineral rights. After that, the remaining owners are forced to participate at a fair market value determined at hearings before the state Corporation Commission.

Other mineral rights remain with the states and with the federal government. Surface and mineral rights on some state lands were set aside to finance education, often in a pattern of one square mile out of every eighteen. That's how the University of Texas in Austin acquired an endowment second only to Harvard's. The federal government controls mineral rights under national forests, Bureau of Land Management lands, and most Indian reservations. National Parks are off-limits to drilling. States control offshore drilling rights out for three miles; farther from the coast the oil rights are federal. Usually, a single oil company nominates a tract of land for bidding. The federal government, at its discretion, conducts a sealed-bid auction, open to all interested bidders, for the drilling rights. In 1953, the first summer I worked for Shell Oil, one of our tasks was to evaluate a recently auctioned tract. Another company had nominated the tract and Shell put in a bid just to keep the other bidders honest. Shell won; we then had to figure out what we bought.

The individual or company that has leased the subsurface rights cannot simply barge in and make a mess of the surface property. The surface property owner has to be compensated for the use of the land.

Pollution with oil or with chemicals is a sensitive problem. One pipeline company asked to lease a two-acre corner of my inherited Oklahoma pasture to build a pumping station. I refused to lease it, because they could close out the lease and walk away, leaving me with the cleanup costs. I offered to sell them the two acres, which they eventually accepted. They never built the pumping station and the cows happily graze there.

When drilling is on state or federal land, damage to the surface is of public concern. A total, blanket don't-drill rule for all federal lands is not an option. We are going to be so hungry for oil that a universal ban on drilling is almost equivalent to committing economic suicide. A balance has to be struck between the expected surface impact and the delivery of oil and gas to the economy. That much we can probably deal with. The big enchilada is the *unexpected* surface impact. Accidents happen. When they happen, they get a huge amount of publicity. I know it sounds like self-serving pleading, but over the years oilfield accidents are becoming less common. The best single piece of news is that the legendary teams of oil well firefighters are slowly vanishing through starvation. Saddam Hussein generated most of their business in Kuwait in 1991 and Iraq in 2003. Don't tell the Sierra Club,[6] but out-of-control offshore accidents are officially recorded on the Internet. Although the reports do not make me want to work on an offshore oil platform, very few incidents involve injuries; some of them release a barrel of oil to the Gulf of Mexico. On the other hand, if the Web site turns you on, head for Morgan City, Louisiana, where they staff the offshore rigs.

Drilling   Procedures

Drilling economics are enormously important. If drilling were cheaper, we could drill higher-risk exploratory wells. Within established fields, we could put in more wells to obtain better recovery. Before 1900, wells were drilled by banging a large chisel up and down on the end of a rope. In 1901, the rotary rig opened the Gulf Coast to drilling. A rotary rig rotates a heavy string of pipe, originally with a toothed roller bit on the bottom. A rather specialized thin mud circulates down inside the drill

The discovery well at Spindletop near Beaumont, Texas, was drilled in 1901.
This photograph, taken in 1902, shows how rapidly the field was developed.
(© 2003 AP/Wide World Photos)

pipe and back up outside the pipe. (The dominant bit was designed by
Howard Hughes's father. A large steady income from Hughes Tool
Company allowed the son to experiment with aircraft, movies, elec-
tronics, and movie stars Jean Harlow, Jane Russell, Ginger Rogers,
Lana Turner, Ava Gardner, Rita Hayworth, Katharine Hepburn, Bette
Davis, Paulette Goddard, Gene Tierney, and Terry Moore.)

I watch eagerly for anything that might lower drilling costs. Lots of
exotic ideas have been tried. Here are some recent events.

- Directional drilling was originally a stunt to reach a location
  that was under a lake or under the Oklahoma State Capitol. As
  techniques improved, wells could be diverted ninety degrees
  from a surface vertical hole to a horizontal hole. Wells could
  follow a single productive horizon for half a mile. When energy
  to rotate the bit can no longer be supplied by rotating the drill
  pipe, the bit is turned by a mud-operated turbine just above the

bit. Saudi Aramco redrilled their largest oilfield, Ghawar, with horizontal holes.

- "Diamond compact" bits make no use of rolling cutters. Sturdy studs protruding from the bottom of the bit are faced with fine-grained synthetic diamonds glued together with tungsten carbide. Sometimes, a single diamond compact bit will drill a well seven thousand feet deep, saving the expense of pulling out the drill pipe to change bits. In 2003, a Japanese group published a technique for making fine-grained diamond compacts cemented with diamond. The new Japanese material is tougher and more heat resistant than the usual material in drill bits.[7]

- Boeing recently donated six patents for high-energy lasers, left over from a discontinued Star Wars program, to the Colorado School of Mines. The hope is to develop a rapid way of drilling oil wells.

- Coil tubing rigs have a reel about thirty feet in diameter. Instead of unscrewing the drill pipe to bring it out of the well, the continuous pipe is rolled up on the big reel like a fishing line. Pipe as big as three and a half inches in diameter can be handled if the reel is large enough. Pipe manufacturers make tapered pipe whose wall thickness slowly increases from the bottom to the top of the hole. My fantasy is that the whole thing will be controlled from an office tower in Houston. The crew on site will consist of a driller and a dog. The driller is there to feed the dog. The dog is there to bite the driller if he touches anything.

As we will see shortly, a New! Improved! drilling technique would not get on the market in time to save us from an oil shortage in this decade. However, fame and fortune could come to the developer of a more economical drilling method. Read the 1968 book *Novel Drilling Techniques* to avoid reinventing the wheel.[8]

Production Methods

Big oil wells initially flow spontaneously. Oil comes to the surface because the pressure of a column of froth—natural gas and oil—inside the pipe is less than the pressure inside the reservoir rock. (Oil squirting over the top of the derrick went out of fashion a hundred years ago.) As production continues, the reservoir pressure falls and the familiar nodding pumpjacks are used to lift oil to the surface.

In early oilfield practice, primary production continued until the wells in a field produced too little to cover the costs. It was then time to institute secondary recovery. This commonly involved pumping water into some of the wells and extracting oil from the others: a waterflood. In some fields, secondary oil almost equaled the amount of primary oil.

Quite good estimates can be made of the total amount of oil that was in the reservoir before production began. It is not guesswork: Well logs obtained from downhole sensors give a measure of the porosity, thickness, and oil percentage in each well. Readings from all the wells in the field, taking each well as typical of its local area, are combined to measure the total oil originally in place. The bad news: Primary production extracts less than a quarter of the original oil. After secondary recovery, about half of the oil is still in the underground reservoir. A long-standing hope is recovering that second half. Unfortunately, after the waterflood, the remaining oil is in isolated droplets in the reservoir rock, and even a fast water flow will not recover additional oil. Modern practice starts the waterflood early, sometimes simultaneously with the initial production.

Especially during the early 1980s, "enhanced" recovery projects sprouted up like bamboo after a spring rain. Detergent floods, fire floods, nitrogen floods, polymer floods, solvent floods, steam floods, and carbon dioxide floods were initiated. More bad news: By the year 2000, two-thirds of the enhanced projects had shut down because of high costs and low oil prices. However, a few out of the list of possible techniques have been successful and the United States produces 750,000 barrels of oil per day from enhanced recovery projects.[9] The big winner has been the injection of carbon dioxide. Although carbon dioxide is soluble in water (seltzer, root beer, Perrier), it is even more soluble in crude oil. The carbon dioxide bulks up the oil droplets and they start to move again. After the oil comes to the surface, the carbon dioxide is stripped off and sent back downstairs to hunt for more oil. Burning some fuels, especially coal, generates carbon dioxide. A major environmental goal is burying the carbon dioxide to avoid global warming. Building a coal-fired electrical-generating plant near an oilfield that needs carbon dioxide is a possible winner. At the moment, carbon dioxide from a natural source in southwestern Colorado is being transported by pipeline hundreds of miles to oilfields in West Texas.

My prediction is that the remaining half of the oil in most reservoirs will not be economically recoverable, even at high oil prices. Lots of cleverness, time, and money have gone into enhanced recovery projects. That doesn't mean that we should stop thinking about enhanced recovery and trying good ideas. It does mean that we'd better not count on using the remaining oil for at least a decade. In fact, it may never be recoverable.

Part of the reason that petroleum is a mature industry comes from reinvesting the large profits over the last fifty years. Worldwide, the major oil and service companies combined spent more than a billion dollars each year on research. In 1960, the budget for the Shell Research Lab in Houston was larger than the total budget of Cal Tech. Truly impressive new techniques were developed; payout from that research helps sustain the industry today. There is a downside: Lots of wheels have already been invented. It is not at all easy, or cheap, to step in at this late date and generate major new innovations.

The Hubbert Outlook

Most of us who predict an imminent decline in world oil production regard M. King Hubbert (1903–1989) as our patron saint. Not that he was so saintly; he was tough to get along with as a scientist. Hubbert wrote half a dozen scientific papers of major importance to geology, but his public fame rests on his controversial prediction in 1956 that U.S. oil production would peak and start to fall in the early 1970s. After his prediction came true, the conservation movement adopted Hubbert as a legend in his own time.

King Hubbert grew up in the central Texas town of San Saba. His bachelor's, master's, and doctoral degrees were earned at the University of Chicago. Before World War II, Hubbert held the rank of instructor at Columbia University for seven years without being promoted to the lowest professorial level. During the war, he worked for the War Production Board, and then he spent thirty years with the Shell Research Lab in Houston. After retiring from Shell, Hubbert taught at Stanford and Berkeley and worked for the U.S. Geological Survey. There is permanent confusion about what to call Hubbert: geologist, geophysicist, petroleum engineer. All are true; it isn't important.

By an odd piece of luck, I started out on good terms with Hubbert. Before I joined the Shell research lab in 1958, I had been a graduate student at Princeton. Harry Hess, the Geology Department chairman at Princeton, was in the National Academy of Sciences with Hubbert. After I moved to Houston, Hess occasionally asked Hubbert whether he had met Deffeyes yet. Hubbert kept answering no. Finally Hubbert rang my office and asked whether I could come to his office for a chat. Hubbert then realized with embarrassment that we had occasionally shared cafeteria tables for more than a year. My last name is so hard to spell that Hubbert never caught on that "Ken" and "Deffeyes" were the same person. He became a close friend for the rest of his life.

Whenever Hubbert grabbed the chalk and started to work through a physics problem, I got the impression that the math was not something he could do effortlessly. A few scientists and mathematicians have an uncanny knack for solving difficult mathematical problems quickly and seemingly with little effort. However, that talent is not universally distributed. As an example at the very highest level, during the years Einstein spent groping for the mathematical formulation for general relativity, he needed, and got, help from a mathematician friend.[10] I'm in the bottom of the ultraslow math category, but I feel a little better knowing that hard work is required even for some high-level scientists.

There are two questions about Hubbert's 1956 prediction of the United States oil peak. Could a valid prediction have been made even earlier? Did Hubbert jump the gun and publish a lucky guess before the data were good enough to justify a prediction? Elizabeth Wood, a distinguished crystallographer (now retired) sent me a letter saying that she knew Hubbert at Columbia University from 1932 to 1935. She reported that he was interested, even then, in limits to the supply of crude oil. Did Hubbert watch for twenty years and pounce as soon as the data became strong enough?

In an attempt to resolve the questions, I fed the historical data one year at a time into a computer program that used Hubbert's equations. For each year, the computer looked back at all the previous years and issued a pseudo-Hubbert prediction, a hindcast. From 1900 through 1933, the computer hindcasts are widely scattered; some of them even predict negative amounts of oil. The predictions do not settle down until after 1958. In my opinion, Hubbert succeeded in 1956 by a combination of real data analysis, shrewd intuition, and good luck. As time progressed, both the data and Hubbert's methods improved. In particular, after 1963 he no longer needed the educated guesses about the total amount of United States oil.

The rationale behind Hubbert's techniques is explained in the next chapter. Some of his successors developed alternative mathematical approaches.[11] I have chosen the cowardly route: stick very close to Hubbert's later methods. What worked for the United States in 1963 has the best chance of making a correct prediction for the world in 2005.

What do the major oil companies know? This is their playground and they can hire all the economists they want. Obviously, they aren't saying much in public. ExxonMobil and ChevronTexaco apparently take a business-as-usual position. BP/Amoco uses "Beyond petroleum" as a motto, complete with a green daisy logo. However, BP's actions seem to be heavily focused on producing oil and selling gasoline. (I was sorely tempted to title this book *Beyond Petroleum* in the hope that BP would give me some free publicity by filing a large lawsuit.) Royal Dutch/Shell announced that they are investing several billion dollars in wind and solar energy projects, and in the spring of 2003 Shell opened a hydrogen filling station in Iceland and a major oil-sand mining operation in Alberta. Total (the former Total/Fina/Elf) is seriously aware of the potential world oil peak. What we do not see is major companies frantically buying up smaller companies to acquire their remaining reserves. In fact, the opposite is happening: The majors are selling off partially depleted oilfields to smaller independents.

There is a long list of excuses, miracles waiting to happen, offered to show the Hubbert analysis to be wrong, incomplete, misguided, or unwelcome. Some combination of access to previously closed areas, resurgence in Russia, and an abundance of new technologies, it is said, will repeal Hubbert's equations. Offsetting these innovations devoutly to be wished are some sure-thing developments on the downside. As an example: Saudi Aramco redrilled Ghawar, the world's largest oilfield, with horizontal holes to improve both the water injection and the oil recovery. It's a success. However, when the water reaches those horizontal recovery wells, there will be a sudden drop in production. Ghawar supplies roughly half of the Saudi oil production, and over a few years there may be a loss of several million barrels per day. It may already have started; one report says that a million barrels per day of water is already coming from the producing wells. This is not a criticism; Saudi Aramco did exactly the right thing at Ghawar. The new efficient horizontal wells, which squeeze out the last producible drop of oil, will let you know rather abruptly when the last drop is gone.

Price Volatility

Some geologists believe, and all economists know, that oil scarcity will result in a new equilibrium price that balances an enhanced search for oil against consumer demands. There is an alternative view suggesting that we will encounter enormous price volatility. The new interpretation arrived as a result of one of my talks. The next morning Suzy Sachs nudged her sleepy husband and announced, "One of Ken's equations is upside down." It turns out that it was right side up for my purpose, but she knew the inverted form from queueing theory. All I knew about queueing was that the word had five consecutive vowels. Suzy had been a systems engineer for the Bell System for twenty years; the phone company knows from queueing. There are several different arrangements for queues: single queue–single server (the grocery store), single queue–multiple server (airline check in), and so on. Suzy explained to me that all of these systems become chaotic when the load approaches the capacity of the system.[12] Customers arrive at random and the time spent waiting in the queue jumps from very short to very long. One of her friends gave me a graph of U.S. natural gas prices. Smooth, steady

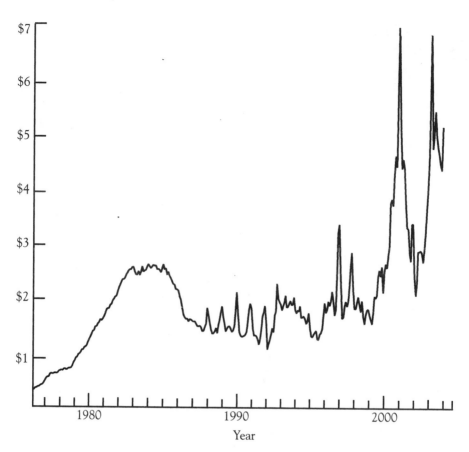

U.S. natural gas prices followed a relatively smooth curve until 1987. After that, increasingly large price swings occurred, especially in winter. This graph shows prices for the first purchase of the gas downstream from the well. There are other, higher prices further along in the gas distribution system.

prices until 1985 and then oscillations get bigger and bigger. Over one weekend during the cold 2002–03 winter, the price of U.S. natural gas doubled.

OPEC was founded to stabilize world oil prices. OPEC monitors a "market basket" of prices for seven different grades of crude oil. From 1983 to 2002, OPEC's goal was to adjust production to hold that average price between $22 and $28 per barrel. (That would be an illegal "conspiracy in restraint of trade" if it were operating in the United

States.) If the price fell below $22, OPEC would cut back production. If the price went above $28, they could flood the market with oil. Since 2002, OPEC has been producing essentially at full capacity. Early in 2004, with oil prices near $33, OPEC announced a *cut* in production. The good news is that OPEC is no longer in charge of the price of oil. The bad news is that nobody is in charge.

Oil price excursions can be triggered by unusual weather, temporary interruptions in the supply system, and swings in the global economic activity. Most oil is used to supply transportation, and on time scales shorter than three years, oil demand is relatively inelastic. On longer time scales, vehicles can be replaced with more energy-efficient transport and we can rearrange our activities to require less travel. Managers *hate* uncertainty, but we are looking into enormous uncertainties in the price of oil. The first time the price swings downward, cornucopians will chant in unison that there was no Hubbert peak. Don't listen.

The threat of lower future prices can put the kibosh on projects that would increase the oil supply. During 1998 and 1999, oil prices (corrected for inflation[13]) fell to the levels previously seen only at the bottom of the Great Depression. An offshore production platform and its wells, or a new tar-sands mine in Alberta, is a multibillion-dollar investment. The return on the investment depends almost entirely on the future price of oil. It's hard to think of a politically acceptable way of reducing the price uncertainty. Possibly the government agrees to make up the difference if oil goes below $22 per barrel, but in exchange the company returns to the government any surplus above $28 per barrel. Some creativity is needed here, both about a fair system and about ways to make the system boll-weevil-proof. One Enron is enough.

Cornucopians

The explanation of the Hubbert methodology in the next chapter is intended to be understandable even by cornucopians. The underlying hypothesis is that the probability of finding oil depends strongly on the fraction of undiscovered oil and that nothing else is of major significance. Of course, one could make a "better," more elaborate model by including the price of oil. But if I knew what oil prices would be over the next two years, I could become obscenely wealthy playing the oil futures market. I wouldn't have to write books or give talks.

It doesn't help to make a list of gifts for Santa Claus to bring to the oil industry. The two most recent development projects for Alberta tar sands have been put on hold. Most of the enhanced recovery methods initiated in the early 1980s are no longer in use. Ethanol from corn requires substantial government subsidies. Investments in research by the oil companies have been cut in half. It's risky to pin your future to a wish list.

A number of economists have stated that the recession from 2001 to 2003 was unlike any seen before. Many of them are focusing on the burst of the dot-com bubble. Most of them have not noticed that world oil production flattened out around 1998. Production will soon be starting a permanent drop; maybe that will get their attention.

Three

The Hubbert Method

In its original form, M. King Hubbert's technique involved some serious mathematics. Many analysts were suspicious that something might be hidden behind the mathematical curtain. Michael Lynch wrote, "For a number of reasons, this work has been nearly impenetrable to many observers, which seems to have lent it an added cachet."[1] This chapter contains a new derivation, using only the simplest beginning algebra, which reaches results identical to Hubbert's. If you absolutely hate math, you can skip this chapter. However, it is an easy opportunity to join the exclusive club of people who actually understand Hubbert.

Even today, Hubbert's methodology is widely misunderstood. Hubbert inadvertently contributed to some of the confusion. His early oil papers have kind of a take-it-or-leave-it flavor. Not until 1982—when he was seventy-nine years old (what I now call "midcareer")—did he publish his reasons for preferring certain formulas.[2] The 1982 paper is pure Hubbert; he never hints whether he knew the 1982 explanation all along or whether he worked out the derivation long after the fact. Also, Hubbert never points out that the graphical method he explained had already been in use for some years by population biologists.[3]

After grinding (slowly) through the 150 pages of the 1982 paper, I finally understood what Hubbert did. Because I am ultraslow with the math, I started working on a derivation that was exactly identical to Hubbert's but with less flash and awe. What follows does not cheat, nothing is left out, and it is the full Hubbert methodology. We'll use the U.S. oil production history as a data set for testing. The United States is very likely the most heavily explored area in the world. It has the world's largest number of actively producing oil wells: 520,000. (China is in second place with 72,000.)

Since we want to estimate likely future trends, nothing beats a straight line on a graph. There are many things that we might choose for the axes on the graph. The simplest choice that I know of, and the one Hubbert used in 1982, plots the cumulative oil production on the horizontal axis and the ratio of annual production to cumulative production on the vertical. (P is for production; Q stands for "qumulative.") The graph of the U.S. production history settles down to a pretty good straight line after 1958. Because the cumulative production always increases with time, time increases from left to right across the graph. It will turn out later that the early points in the upper left-hand corner of the graph occur not because the production was higher than the 1958–03 line. Instead, the production happened when the cumulative production (in the denominator) was low.

Now I, and my computer, want to find the best-fitting straight line through the points for 1958 through 2003. I warn you, when you let me draw that straight line, you will have bought into the whole Hubbert story lock, stock, and barrels. The rest of this analysis is nothing more than examining the implications of that straight line.

The first thing to notice is the place where the straight line meets the horizontal axis: 228 billion barrels. In one sense, it is only a number describing the momentary position of the line. In the Hubbert interpretation, that intercept is the expected amount of oil extracted from the United States when the last well finally runs dry. Let's call it Q_t for the cumulative total. In contrast, the U.S. Geological Survey in 2000 estimated a U.S. total of 362 billion barrels.[4] Either that straight line is going to make a sudden turn, or the USGS was counting on bringing in Iraq as the fifty-first state.

The intercept of the line with the vertical axis, which we will call

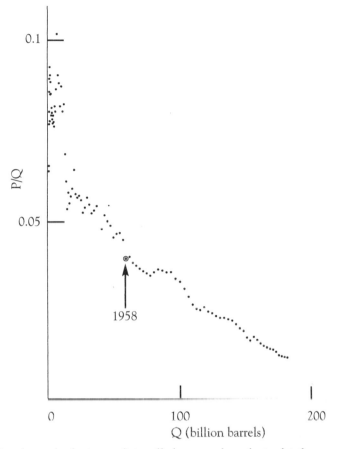

Population biologists traditionally have used graphs in this format to examine population growth. Hubbert showed that the same graph could be used to analyze oil production. Each dot corresponds to one year of U.S. oil production. The horizontal axis is the cumulative production (Q) from 1859 up to a particular year. The vertical axis is the production in a given year (P) divided by the cumulative production up to that year (P/Q). From 1958 onward, the points seem to line up fairly close to a straight line. Many of the points before 1958 lie above the trend—not because production was too high, but because it happened too early. When Q is small, P/Q becomes large.

a, is sort of an interest rate. It's the annual production expressed as a fraction of the cumulative production. On the U.S. production graph, the vertical intercept gives 0.0536 for a, equal to 5.36 percent per year.

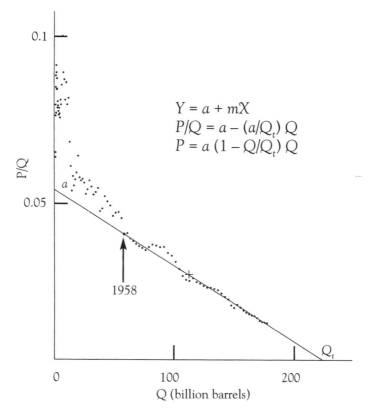

$$Y = a + mX$$
$$P/Q = a - (a/Q_t) Q$$
$$P = a (1 - Q/Q_t) Q$$

The straight line that best fits the data from 1958 to 2003 in the previous graph is drawn in here. All of Hubbert's theory follows from drawing that straight line. No further assumptions or guesses are needed; that line *is* the Hubbert theory. Where the line crosses the vertical axis (labeled a) is an idealized beginning, when the cumulative production for the United States was zero. The straight line reaches the horizontal axis (Q_t) when the annual production, P, falls to zero. The uppermost equation is just the equation of a straight line. Y is the vertical axis of the graph and X is the horizontal axis. The symbol a is the value of Y when X is zero, and m is the slope of the line. The middle equation is simply a translation, putting in the symbols from the graph. Y becomes P/Q, X is Q, a keeps the same meaning, and the slope of the line is a/Q_t. The minus sign in the middle equation arises because Y is decreasing as X increases. Multiplying both sides of the middle equation by Q produces the bottom equation. This is an algebraic statement of the Hubbert theory. The key concept is inside the parentheses, which can be written either as $1 - Q/Q_t$ or equivalently as $(Q_t - Q)/Q_t$. In either form, it is the fraction of the total oil that remains to be produced. The bottom equation describes a bell-shaped curve, specifically a "logistic" curve. The peak of the production rate occurs when half of the oil has been produced. A small + sign on the graph indicates that midpoint.

Inside the parentheses in the lowermost equation is Hubbert's heavy magic. Q/Q_t is the fraction of the total oil that we have already produced and $1-Q/Q_t$ is the fraction yet to be produced. That equation says that our ability to produce, P, is linearly dependent on the fraction of oil that remains. When Drake drilled his first well, Q was 0 and the term inside the parentheses was equal to 1. At the very end, Q becomes equal to Q_t and $1 - 1 = 0$; we don't produce any more oil. In between, it slides down a straight line from 0 to 1. Later, we will use this same equation for finding oil. The idea being tested is whether the ability to find oil is dominated by the fraction of oil that remains undiscovered. The ease of catching fish depends mostly on how many fish remain in the pond.

When the term inside the parentheses is equal to 1, or if the whole term—parentheses and all—were taken away, the equation describes unlimited compound-interest (exponential) growth. Unbounded exponential growth warms the hearts of cornucopians. The entire debate between Hubbertians and cornucopians comes down to what is inside those parentheses.

A long-standing tradition in science, dating back before 1350, suggests trying the simplest hypothesis first (called Occam's razor). If the shoe fits, wear it. Don't invent elaborate schemes if the simplest explanation fits. Postulating that our ability to produce oil depends entirely, and linearly, on the unproduced fraction is the simplest idea that could be tested. It fits the 1958–03 U.S. production data reasonably well. What infuriates cornucopians is Hubbert's implication that nothing else matters, only the undiscovered fraction. What about 3-D seismic, deeper-water drilling, ANWR, computer imaging, and increased oil and gas prices? M. King Hubbert is not available to reply. On his behalf, I will say that improved technologies and incentives have been appearing all along, and there seems to be no abrupt dramatic improvement that will put an immediate bend in the straight line.

We can easily compute $1/P$, which comes out in years per billion barrels. Since 1933, the United States has produced more than a billion barrels per year, so for those years $1/P$ is a fraction of a year. For each billion barrels of cumulative production, Q, we use a (0.0536) and Q_t (228.4) to compute a time interval, $1/P$. As we go along, we keep adding up the time intervals to get the total time. The resulting graph is a bell-shaped curve.

The bell-shaped curve is symmetrical in time; the downside is a mirror image of the upside. This is nothing more than the consequence of the linear dependence of the production rate on the unproduced fraction. The curve is called a "logistic" curve, but the name needs explaining. "Logistic" and "logistics" are two different words that evolved from the same root, the Greek word for "calculating." "Logistics" is not the plural of "logistic." "Logistics" today refers to keeping an army lodged, fed, and transported (for which you need to calculate how many, how much, for how long). "Logistic" means "skilled at calculating." Pierre François Verhulst, a Belgian who developed the logistic curve in 1838, was skilled at calculating.

The area under the logistic curve is Q_t. Because of the symmetry in time, the peak year occurs when the area under the curve reaches half of Q_t. The maximum production during the peak year is $0.25aQ_t$. The annual production falls to half of the maximum at $1.762 / a$ years before and after the peak.[5]

Wait a minute. We haven't labeled the years on the curve.[6] We absolutely do not want to deal with the starting point of the logistic curve. No telling how many barrels of oil were scooped up out of oil seeps by my Native American ancestors before Drake drilled his well. What we do is choose, from the data, a year or a year and a fraction, corresponding to some integer value of the cumulative production. For instance, the U.S. cumulative production equaled 100 billion barrels in 1972.1. We then run the $1/P$ time summation in one-billion-barrel steps until we get to 100 billion. At that point, we set the "clock" to 1972.1. Then we can go back to the beginning and start the calculation all over, knowing that the cumulative production will go through 100 in 1972.1. Now we can generate a graph comparing the actual U.S. production to the smooth logistic curve. Not bad, actually.

The logistic curve peaks in 1976, but the actual year of greatest U.S. oil production was 1970. We will see later that something similar may be happening with world production: a logistic peak in 2005 and the single greatest year might be 2003. Why does a local peak happen before the smoothed peak? It certainly can't happen after the peak, because diminished capability is holding the production down. If it happens at all, it has to happen before the smoothed peak, when there is unused capacity.

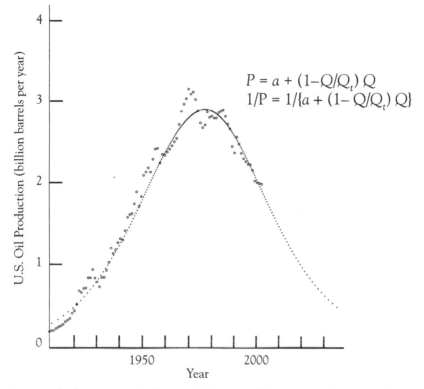

$$P = a + (1-Q/Q_t)\,Q$$
$$1/P = 1/\{a + (1-Q/Q_t)\,Q\}$$

Let's get back to the graph that we all know and love: production versus time. The top equation is repeated from the previous graph. Take the reciprocal of both sides, which is the same as dividing 1 by both sides of the equation, to get the bottom equation. We now have years per billion barrels instead of billion barrels per year. The computer goes along, adding up the years and fractions of years for each billion barrels. Each small dot is one billion barrels of production, corresponding to the best-fitting straight line on the earlier graphs. The open circles show the actual history.

I prefer to avoid making strong analogies between the biological origins of the logistic curve and Hubbert's borrowing. In the most primitive sense, the logistic equation is a bit of algebra that gives a reasonable fit to the oil production history. If it didn't fit, we wouldn't be talking about it. Because it does fit, the possibility arises that the un-produced fraction of the total oil dominates over all other factors. Of course, the price of oil matters; it just doesn't matter very much.

For the biological model, imagine a bulldozed vacant lot starting to fill up with weeds. In the beginning the weeds are far apart, they are not competing with each other for resources, and the number of weed plants grows like compound interest. As the weeds become more numerous, the rate at which new weed plants appear starts to depend on the unoccupied space in the vacant lot. Once the lot is fully occupied, we still have weeds, but the number of plants remains constant.

In the biological situation, growth is the number of births minus the number of deaths. In the oil analogy, no barrel of oil ever dies. Although barrels of oil don't have babies, oil discoveries *do* have intellectual offspring. The discovery of a new "play"—for instance, reefs in Alberta—begets a whole string of similar discoveries. What the two models have in common:

- The growth rate of a biological population depends linearly on the unoccupied fraction of the environmental carrying capacity.
- The oil production rate depends linearly on the fraction of the total oil that remains to be produced.

The World Picture

Now that we know how the Hubbert method works, we're ready to take on the world. In round numbers, world oil events take place about thirty years after the U.S. equivalents. The United States can import more than half of its oil precisely because the world picture is at a less mature stage. We begin, as before, by graphing production (P) and cumulative production (Q). On the graph of P/Q versus Q, world production settles down to a reasonably straight line after 1983, corresponding to 1958 for the United States.

The best computer-fitted straight line to the world production data from 1983 through 2003 extends downward to an estimate of two trillion barrels for Q_t, the total world cumulative production when the last dog is dead. Many petroleum analysts predict eventual world production limits within 10 percent of that estimate. Even more analysts, cornucopians by definition, accept much larger estimates.[7] As an example,

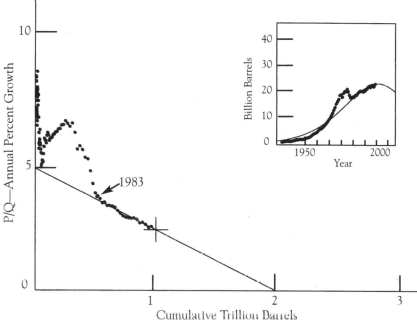

World oil production settles down after 1983 to a reasonably straight line. The best-fitting line to the interval from 1983 to 2003 has an *a* intercept of 0.05 and a Q_t of 2 trillion barrels. The dots are one year apart. A plus sign marks the time of maximum production, when half of the oil is gone. My estimate of the peak date of the smooth curve is November 24, 2005, Thanksgiving Day. The uncertainty is roughly one month on either side of Thanksgiving. Of course, production in individual years bounces up and down a little so it is not at all clear which calendar year goes into the record book. The inset graph shows the same data as production versus time.

the U.S. Geological Survey put one hundred person-years into estimating the world oil potential and came up with 3.012 trillion barrels. None of us has two hundred person-years available to dissect and to criticize the USGS analysis. What seems to have happened is illustrated by the story of Shell interviewing as potential employees a geologist, a geophysicist, and a petroleum engineer (the kind that estimates reserves). One question asked was, "What is two times two?" The geologist answered that it was probably more than three and less than five, but the issue could use some more research. The geophysicist punched it into his palmtop computer and announced that it was 3.999999. The petroleum engineer jumped up and locked the door, closed the window blinds, unplugged the phone, and asked quietly, "What do you want it to be?"[8]

Even more important is the tick mark on the line when half of Q_t has been produced. That's the peak of the best-fitting logistic curve. That's the top of Hubbert's peak. That's the year I have been dreading, and it's essentially here. We extend the straight line trend by only a smidgeon to get to the peak. Even if some future miraculous technology were to bend the line out to three trillion barrels, we need that miracle in place this year. There is every indication that the world will ride over the peak and then ask what happened.

The single greatest year of U.S. oil production was 1970, not the 1976 peak of the smooth curve. Even those of us concerned with the problem did not notice right away; it was entirely possible that some later year might exceed the U.S. 1970 production. Confirmation came when the Texas Railroad Commission in 1972 ended production rationing in Texas. (Most oil was originally shipped on railroad tank cars; during the Great Depression, the Texas Railroad Commission acquired the task of supporting the national price of oil by limiting production in Texas.) The end of production rationing indicated that the United States was fresh out of unused production capacity.

OPEC was founded after a close study of the Texas Railroad Commission. The Texas experience showed that regulating production from one dominant producer could set the crude oil price for the entire industry. OPEC, founded in 1960, started with five major producing countries as members. Eight other members joined later. Through the 1990s, OPEC members, one by one, ran out of surplus production capacity until Saudi Arabia was the last man standing. From roughly 1997 onward, the Saudis were uniquely able to regulate production. The aim was to hold the world oil price in the OPEC target range from $22 to $28 per barrel. In a little-noticed news item, on March 6, 2003, the Saudi government announced that their production had maxed out at 9.2 or 9.5 million barrels per day.[9] As of 2003, no significant underutilized oil production capacity existed anywhere in the world.

This is an enormously important conclusion. So far, we have examined three clues:

- The graph on page 43 says that the smooth logistic curve will peak in 2005, or the first months of 2006.
- World crude oil production has been almost flat since 1998.
- No country, including Saudi Arabia, has unused production capacity.

Before we seize on those three clues as strong enough to guide policy, some additional data are relevant.

Production is the last of three steps in the "upstream" supply chain. The first step is finding the oilfield. This happens only once; after you drill the first successful well, you won't forget where it was and discover the field again. Over time, the field may get larger or deeper, but it is inevitable that the full extent of the field will eventually be tapped. One minor point: Occasionally several independent oilfields grow and eventually merge into a single giant, but only the earliest well in the combined field actually counts.

After production begins in an oilfield, development drilling expands the geographic extent—and sometimes the depth—of the field. This process can take decades. As wells are placed in extended portions of the reservoir (or reservoirs), the oil likely to be produced from those actual wells is listed as "reserves." The sum of the individual-field reserves is supposed to add up to the reserves for the country, and eventually to the world total.

Production Plus Reserves

Since we started with production, we will work backward and deal with reserves and then examine the date of the first successful well in each field. In his 1962 paper, Hubbert used the U.S. reserves to avoid using "expert" opinions on the total recoverable oil. Hubbert used the word "discovery" to mean all the cumulative oil produced in a given year *plus* the known reserves as of that same year. Hubbert's "discovery" is sort of points on the scoreboard: all the oil found up through that year. I had a terrible time with the preceding sentences. My fingers wanted to type "discovery" in the usual sense of the first well in the field. Because there is already an enormous amount of confusion about Hubbert's methods, I will use "discovery" for production plus reserves and use "hit" when I mean the first well in a field.

Reserve estimates are messy. In the cleanest sense, reserves are simply length × width × thickness × porosity × expected percent recovery. Usually, there is some small-scale fudging because most engineers are conservative and prefer their estimates to be a little bit low. On a large

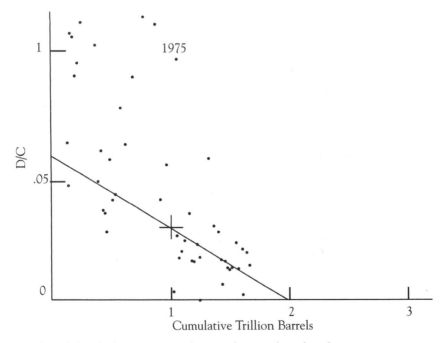

Hubbert defined "discoveries" as the cumulative oil produced up to a given year *plus* the reserves known in the ground as of that same year. On this graph, the abrupt reserve increases reported in most OPEC nations during the 1980s have been subtracted out. Also, at the end of 2002, *Oil & Gas Journal* abruptly added 175 billion barrels of reserves in the Canadian tar sands, which are not included here because production facilities at that scale are not in place. The most recent giant oil discovery was Cantarell, in offshore Mexico, in 1975. This graph shows no reason for rejecting the previous estimate of 2 trillion barrels for the ultimate recovery.

scale, political priorities enter. (When I was an undergraduate, we called it "multiplying by Finagle's constant.") The Hubbertian geologist Colin Campbell identified the largest instance of political reserves: In the mid-1980s, the oil reserves of most OPEC countries abruptly doubled.[10] No new wells were drilled, but each country's share of the OPEC production quota had been revised to include reserves as well as production capacity. Campbell (and I) divided by Finagle's constant to discount the abrupt increases. Initially, I was too harsh. In retrospect, about 20 percent of the abrupt OPEC increases seem to have been genuine revisions of reserves and 80 percent materialized out of thin air.

It is possible for discoveries, defined Hubbert's way, for a year to be negative. If the downgrading of reserve estimates in a year is greater than the actual production for that year, discoveries have a minus sign.

The methodology for analyzing discoveries is the same as for production. Plot cumulative discoveries horizontally, and on the vertical axis plot annual discoveries divided by cumulative discoveries. Even for the most recent years, the points have greater scatter than those we saw earlier on the production plot, and the points after 1985 depend on handling the abrupt increase in announced OPEC reserves. However, a computer-fitted line says that we passed the peak of world discoveries back in 1977. Even more striking is the prediction that the oil we have now "discovered" (in the sense of being recognized as reserves) is 82 percent of all the oil we are ever going to discover. Gloom.

Hitting New Oilfields

As mentioned above, "discover" is a Hubbertian word, so I will refer to the initial well in an oilfield as a "hit." Hit years for oilfields are generally known; Carmalt and St. John published a world list of all oil and gas fields larger than 100 million barrels of producible oil.[11] We are *not* counting how many oilfields; we are counting how many barrels are eventually expected from the total hits for each year. Same old method: Plot cumulative hits horizontally and the ratio of annual hits to cumulative hits vertically. Draw the best-fitting line. Total expected hits, when the hittin' is done: 2.013 trillion barrels. Middle year and peak of

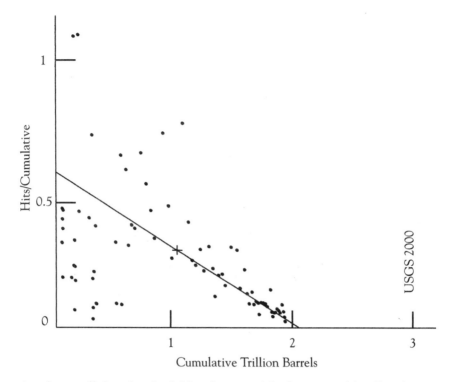

Attributing all the oil in the field to the year of the first successful well in that field produces this striking conclusion: The oilfields already in production contain 94 percent of all the oil we are ever going to find. Instead of an ultimate production of 2 trillion barrels, the best-fitting line on this graph hits the horizontal axis at 2.013 trillion. This is far below the 3.012-trillion-barrel estimate from the U.S. Geological Survey's 2000 *World Petroleum Assessment.* Four Middle East oilfields are large enough to drive their discovery years off the graph (and off the top of the page), although they are not excluded from the statistics. Those fields are 1927 Kirkuk, Iraq; 1928 Gach Saran and Haft Kel, Iran; 1938 Burgan, Kuwait; and 1948 Ghawar, Saudi Arabia. On occasion, several apparently independent oilfields merge into a single giant oilfield. Deleting all the duplicates in the list is a slow painstaking task. Also, the U.S. Gulf Coast and offshore West Africa contain significant oil but in large numbers of less-than-giant oilfields that are not as easy to track.

the maximum smooth logistic curve: 1965. But worst of all, we have hit 94 percent of all the oil that we can ever expect to hit. Doom.

All Together Now

We can put together a composite picture of the world oil situation by coaxing production, discoveries, and hits onto one graph. Here is the numerical scoreboard for the best-fitting lines:

	Years Used	a	Q_t	Peak	Percentage
Production	1983–2003	0.059	2.013	2005	49
Discoveries	1976–2002	0.072	2.013	1978	82
Hits	1965–2003	0.081	2.013	1964	94

Here I make a modest break with the rationale that Hubbert used. Hubbert treated "discoveries" as a time-displaced replica of the production curve. That cannot be the case. On the morning of August 28, 1859, Edwin Drake and his driller found their hole filling with oil. Drake did *not* say to the driller, "We can start producing twenty-two years from now." World production, discoveries, and hits started off all together. Hits initially have to grow faster than discoveries. Discoveries initially have to grow faster than production.

The hits graph certainly looks as if the oil game is almost over (see page 50). The fat lady is singing. "Only" 100 billion barrels left to find. Anyone who claims 60 billion barrels for the Caspian Sea eats up most of the leftovers. My guess is that the remaining discoveries will look more like change for a hundred-dollar bill: one twenty, two tens, two fives, ten ones, a roll of quarters, two rolls of dimes, five rolls of nickels, and twenty rolls of pennies. Don't sneer. Each of those pennies corresponds to 10 million barrels of produced oil: at $40 per barrel, that comes to $400 million.

Is this the inevitable story? In one sense, back on page 38, if you let me draw that straight line through the U.S. production history, you bought into the entire Hubbert analysis. After drawing that straight

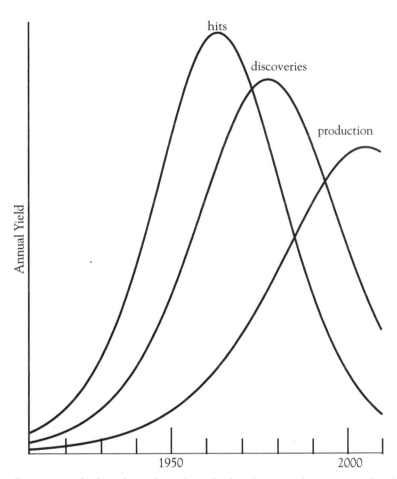

hits

discoveries

production

Annual Yield

1950

2000

Converting the best-fitting lines from the last three graphs into annual yield versus time illustrates a mistake made by Hubbert. Hubbert stated that his discovery curve would be identical in shape to the production curve but displaced by a fixed number of years. The areas under the three curves are identical: 2.013 trillion barrels. The one that develops fastest (hits) has the highest and narrowest peak. Discoveries, used in Hubbert's sense of cumulative production plus reserves, occupy the intermediate position between hits and production.

line, no further theoretical guesses or approximations were made. (Well, I did have to unfudge the OPEC reserve increases.) The major theoretical conclusion says that a straight line requires that production (discoveries, hits) depends linearly on the fraction of the oil that is un-produced (undiscovered, unhit).

In the particular narrow instance of world oil production through the peak years, the line is extended for only a short distance. Even if some miracle were to happen later to move the total world production beyond 2 trillion barrels, that miracle can't turn production sharply up in the next couple of years.

Four

Mostly Gas

If you look at a drilling rig, you can't tell whether it's searching for natural gas or for oil. Not only do the drill rigs look the same, but the exploration strategies and the well completion techniques are also similar. Natural gas, like oil, is lighter than water so the strategy is to locate some kind of concave-downward trap that would hold oil or gas and prevent it from floating all the way to the surface. (To conserve paper, in this chapter "gas" always refers to natural gas, not to gasoline.)

Two different units, which are almost identical, are used for measuring and selling natural gas. One unit is a volume of 1,000 cubic feet at room temperature and pressure. The other is a heat generation of a million BTUs when burned. (BTU is a British Thermal Unit, enough heat to raise the temperature of one pound of water by one degree Fahrenheit.) Burning 1,000 cubic feet of ordinary natural gas generates close to a million BTUs. If anyone is having metric withdrawal symptoms, a million BTUs is close to a gigajoule. Unless you are splitting hairs over a major gas-delivery contract, the units are practically identical.

Despite their similarities, there are major differences between gas wells and oil wells:

- Producing gas wells do not have pumpjacks happily bobbing their horseheads up and down while pumping oil up out of the well. At the top of a gas well is a collection of pipes and valves, known as a Christmas tree.
- Virtually all gas is moved in pipelines. The cost of a pipeline connection is a major part of natural gas economics. Large gas fields justify building new pipelines, but a small isolated field with one or a few wells would never pay back the cost of a pipeline connection.
- Natural gas flows out of the rock into the well with much less friction than oil or water. In technical terms, the viscosity (resistance to flow) of gas is about a factor of one hundred smaller than water or than the lightest grades of crude oil. Rocks with permeability too low to produce economic amounts of oil can be profitable for natural gas.

Natural gas comes from several different underground settings. The first distinction is between conventional and unconventional gas. "Conventional" gas refers to the types of deposits that have historically supplied most of the world's gas. "Unconventional" (wouldn't you guess) is used for categories that are less thoroughly explored. However, none of the unconventional types are intellectually new; examples of all of them have been known for thirty or more years. I discuss unconventional gas later in this chapter. Here are the conventional gas sources:

- Solution gas: Underground oil and the gas dissolved in it separate into a froth of gas and oil as the pressure drops. Solution gas is what drove the historic oil gushers that blew oil over the top of the derrick. However, as the pressure inside the oil reservoir drops, the resulting froth inhibits the flow of oil to the bottom of the well. Oil production driven by solution gas is notorious for recovering only 20 to 40 percent of the oil in place. Further, since there is no way to hold back solution gas and produce only the oil, the gas is essentially a free by-product of oil production. Until about 1965, solution gas production kept the price of natural gas down around 3¢ to 7¢ per 1,000 cubic feet. The price as of 2004 was $6 per 1,000 cubic feet.

Gas wells normally do not have pumps or permanent derricks. This gas well in western Siberia is typical. The percentage of the world's natural gas in western Siberia is almost as large as the percentage of the world's oil in the Middle East.
(© Bryan and Cherry Alexander)

- Gas caps: The porous rock in some underground reservoirs contains more gas than the oil can dissolve so a separate "gas cap" forms above the oil. It is possible to produce the oil and save the gas cap for future production. The option exists to drill the well all the way through both the gas- and the oil-bearing layers and to cement in a steel liner (called casing) right to the bottom of the well.[1] Downhole wireline logging tools measure the depths to the gas, oil, and water. Shaped explosive charges perforate holes through the casing and open only the oil-bearing layer for production. An expanding gas cap is somewhat more efficient at recovering oil: 30 to 50 percent of the oil originally in place is recovered. After the more profitable oil is produced, the gas in the gas cap is still available.

- Beyond the oil window: The "oil window" is a zone from 7,500 to 15,000 feet. At 7,500 feet, temperatures are usually hot enough to start cracking larger organic molecules into liquid oil with five to twenty carbon atoms per molecule. The cracking process continues the deeper you go. At 15,000 feet, the oil is cracked down to molecules with only one carbon atom: methane, the dominant component of natural gas. Rocks that are now, or have ever been, buried below 15,000 feet are candidates for gas production but not oil production. (There is a minor exception: Because of very rapid sedimentation or abnormally low temperatures, a few locations have oil down to 17,000 feet.) Oilmen learned to avoid these gas-only areas. However, after natural gas prices started to rise above $1 in 1979, these oil-free areas became economic candidates for drilling.

At the moment, in the United States, drilling for natural gas has eclipsed drilling for oil. The Baker Hughes count of U.S. active drilling rigs for March 19, 2004, showed 162 rigs drilling for oil and 963 looking for gas.

When will we run out of natural gas? It is not an easy question. The analysis that M. King Hubbert devised for crude oil is not readily adapted to natural gas. Although the *Oil & Gas Journal* end-of-year reports give natural gas reserves for each country, they do not include natural gas production. There are several explanations. Some natural gas is used locally and never metered, whereas crude oil gets monitored once when it is produced and again when it enters a refinery. An additional problem concerns gas associated with oil production that is "flared," deliberately burned in a huge, controlled open-air flame. Is that flared gas to be counted as production in the Hubbert sense?

Fifty years ago, much of the natural gas produced with oil was flared. I vividly remember driving past a gas flare at Fiddler Creek, Wyoming, in 1952. The heat was incredible: Over a hundred-yard radius the sagebrush had been roasted. When the first astronauts orbited Earth, the most prominent visible human effect was the gas flares in North Africa and the Middle East. Given the escalating price of gas, it comes as a surprise that any gas at all is still being wasted. (Just venting the gas to the air without burning it is even worse: Methane is much more powerful as a greenhouse gas than carbon dioxide.)

Here is the scoreboard for flared gas, in billions of cubic feet per day, as reported by the U.S. Energy Information Agency:

Historically, where natural gas was produced along with oil, much of the gas was burned in open-air flares. This is an offshore gas flare, near the Sinai peninsula. (Kevin Fleming, CORBIS)

3.7	Africa (1.7 is in Nigeria)
1.7–3.2	former Soviet Union
1.6	Middle East (this number seems suspiciously low)
0.7–2.0	Asia
1.2–1.7	North America
0.3	Europe
9.2–12.5	Total world

For comparison, the United States consumes an average of 60 billion cubic feet per day. Not all gas is consumed or flared. Some is injected back into the top of the oil-saturated layer to build an artificial gas cap or to enlarge an existing gas cap. (This is the current practice in the giant Prudhoe Bay field in northern Alaska.) The gas injection improves oil recovery, and the gas is preserved for future marketing.

Natural gas that makes it into the pipeline serves a variety of markets. Within the United States, gas is used for electric power generation (24%), industrial use (35%), commercial building heat (16%), and residential heating (25%). "Industrial" includes some nonheating uses. Some goes into petrochemicals. Hydrogen derived from natural gas is reacted with atmospheric nitrogen to produce fertilizer.

Using 24 percent of the natural gas for generating electricity is a new phenomenon. During the Nixon administration, natural gas was recognized as a premium fuel, and generating electricity with gas was absolutely forbidden. Here is the scoreboard for the year 2000:

Annual U.S. Electricity Generation (billions of kilowatt-hours)

coal	1,968
nuclear	752
natural gas	612
hydroelectric	273
oil	109
other (including wind and geothermal)	84

There is a division of labor among the electric power sources. Nuclear power plants operate best at a steady load, twenty-four hours per day. Therefore, the base load, the demand that exists all the time, is carried mostly by nuclear plants. Nuclear power plants also typically have very high capital costs and low operating costs. Coal-fired plants, with their low fuel costs, dominate the intermediate market for demands beyond the base load that exist through part of the day. At the few times of extreme loads, as arises on a miserably hot and humid summer afternoon, what is needed is power-generating capability with low capital cost, even if it means higher fuel costs. The answer is a gas turbine, similar to a jet aircraft engine but burning natural gas and driving an electric generator.

The increased, post-Nixon use of natural gas for electricity generation has used up the surplus gas-producing capacity for North America. The potential victim is the natural gas–powered automobile. As of 1981, natural gas looked attractive as a motor fuel. It puts less carbon dioxide in the atmosphere than ordinary gasoline, it has an octane rating of 135, and the inside of the engine stays clean. Natural gas filling stations were already in service in Italy, New Zealand, and British Columbia in 1981. Natural gas producers and distributors loved the idea because peak driving happens in the summer, whereas gas consumption for heating residential and commercial buildings peaks during the winter.

Natural gas vehicles using less than the equivalent of ten gallons of gasoline per day usually carry the gas in high-pressure tanks. Consequently, natural gas filling stations require a fair amount of special equipment and there arises the chicken-egg problem. Individuals will not buy natural gas–powered vehicles because there are no filling stations; filling stations don't exist because nobody owns natural gas vehicles. In an interesting new development, Honda is planning to market a home gas compressor that sits in the garage and fills the high-pressure tanks overnight. Most U.S. homes have existing gas connections for cooking or space heating. Commuters would find it attractive because the gas costs less than the equivalent amount of gasoline. Almost everyone would find it attractive because there will be no obvious way for the government to collect taxes on the fuel. Avoiding taxes is America's second-favorite indoor sport.

The unusually cold winter of 2002–03 gave us a look at the future of the natural gas market. In early March 2003, continued high demand caused the stored reserves of natural gas to fall below the trillion-cubic-foot level, the equivalent of a "low fuel" warning in a car. Pipelines shut off gas to those customers with interruptible contracts. Several analysts were writing that North American gas production was declining, although there was little agreement on the rate of decline. Since another winter like 2002–03 will eventually happen again, we have no immediate flexibility to divert major amounts of natural gas to automotive use. In the future, the only hope for natural gas automobiles would be converting more of our electricity generation to nuclear or to coal plants, thereby freeing natural gas for automotive use.

In energy content, it takes about 6,000 cubic feet of natural gas to equal a barrel of oil. Since most natural gas is almost pure methane (one carbon, four hydrogen atoms), the energy content from different areas is almost identical. The only significant exception comes from so-lution gas associated with oil production, which usually contains some

A compressor for at-home fueling of a natural gas powered car was introduced in 2002 by FuelMaker Corp, owned in part by American Honda. In the first year, ten thousand units were sold, at an average price of approximately $1,000. (Joe Marquette, Associated Press)

hydrocarbon gases with two or more carbon atoms. It doesn't sound like a big deal, but one producer in northwestern Pennsylvania put some oil-associated gas in the pipeline, and a customer farther down the pipeline melted an industrial furnace.

Most of the time, the heat content determines the prices of different fossil fuels:

Fossil Fuel Prices, March 22, 2004

Fuel and Location	Price (per million BTUs)
crude oil, West Texas Intermediate	$6.375
residual bunker fuel, New York harbor	4.033
heating oil, New York harbor	6.733
propane, Mont Belvieu, Texas	6.441
ethane, Mont Belvieu, Texas	6.259
natural gas, Henry Hub, Louisiana	5.960
coal, Big Sandy, Kentucky	2.525

These are called benchmark prices for a standard grade of fuel available at a recognized marketplace. There are two kinds of prices: a spot price for an immediate cash sale (as in the table above) and futures prices for contracts to deliver on specific dates. On the futures market, contracts for standard products from oil to pork bellies are bought and sold like stocks and bonds. In addition, major consumers and major producers often arrange private contracts. The spot market for natural gas is at Henry Hub, a pipeline junction in southern Louisiana about midway between New Orleans and Beaumont—deep Cajun country. Spot markets sometimes seem to be in illogical places. The U.S. spot market for baled hay is at Petaluma, north of San Francisco, near the coast. The world spot market for crude oil is Rotterdam harbor in the Netherlands. In the spot-price table, coal costs about half as much for the same heating value. If it were not for the pollution-control problems, coal could partially displace other fuels.

Most gas is not traded for cash at the Henry Hub spot market; more than half is sold through contracts. Publicly traded, standardized futures contracts are agreements to buy or sell ten million BTUs of natural gas, at Henry Hub, on upcoming dates a month apart. Gas consumers and producers use these contracts to lock in a definite price. In addition, there are publicly traded options to buy or sell futures contracts. Speculators love it; there are complicated trading strategies with names like bull spread, iron butterfly, and short strangle. In a less colorful mode, gas producers and consumers—especially utilities—work out specific nonstandard, nontradable contracts.

Although natural gas and liquid crude oil sound like very different materials, there is no clean dividing line between them. The most familiar intermediate product is stored in tanks under backyard gas grills and on the back of travel trailers, known as bottle gas or liquefied petroleum gas. Propane and butane are liquids under the pressure inside the tank. When the valve is opened, the liquid starts converting to vapor, and the vapor is burned much like natural gas.

The spot marker for buying and selling natural gas, and the basis for natural gas futures contracts, is a pipeline junction at Henry Hub, Louisiana. (Lowell Georgia, CORBIS)

Near the bottom of the oil window, the predominant hydrocarbon molecules have four to six carbon atoms. That contrasts with one carbon in methane gas and six to thirty carbon atoms in the various molecules in oil. These intermediate products are most often referred to as natural gas liquids or gas condensates. Some statistics include them in with crude oil production, some treat them as a separate category, occasionally they go unreported.

Before 1971, when the Texas Railroad Commission was rationing Texas oil production, natural gas liquids were not considered to be oil. Huge amounts of money flowed from that simple distinction. In 1962, I was shown samples of reservoir rock from a gas condensate field in East Texas. The rock was so porous that it looked like lacework. I was surprised that it was strong enough to hold up the overlying rock layers. Because there was no production rationing on natural gas liquids, the oil company explained that the field was not a cash cow, it was a cash river.

There have been newspaper and magazine stories over the last twenty years discussing the possibility that natural gas might have come from inorganic sources deep inside Earth. During the 1970s, Professor Thomas Gold, a Cornell astrophysicist, championed the inorganic origin for methane, and in 1986, he convinced the Swedish government to drill a deep hole beneath a fossil meteorite strike. No gas. Lately, J. F. Kenney in Houston, along with three Russian colleagues, has been claiming that gas and even oil have inorganic origins.[2] If these suggestions were true, there could be important sources of natural gas in places where petroleum geologists usually do not look.

It seems utterly obvious to me that natural gas is the end product of the thermal cracking process that generates oil. There are molecules in oil with easily recognized resemblances to organic material, sometimes called molecular fossils. For instance, porphyrins in crude oil are very similar to chlorophyll in plants. Because methane is such a small, simple molecule, there is no way to apply the molecular fossil approach to natural gas. This left a window open for the claim that natural gas was an inorganic product migrating upward from the earth's interior. An enormously useful study, which received little attention at the time, examined gases that likely had migrated upward from deeper in the earth. The deep copper mine at Kidd Creek, Ontario, is in very old and very hard rocks. The carbon isotope signature of the mine gases had no resemblance to the natural gas from commercial producing wells.[3] To me, the Kidd Creek observations say that there is no hope of finding abundant supplies of natural gas in deep hard rocks.

Claims like Gold's and Kenney's present a dilemma for scientists. Science has to be open to new ideas; major scientific revolutions grow from roots that initially seem implausible. Yet for every unconventional idea that grows, there are thousands of proposals that sound kooky because they are kooky. Working scientists cannot spend all of their time entertaining kooks. Unfortunately, some of the kooks scream that we are afraid to consider the possibility that we are wrong and they are right. We get our ordinary work done in part because a few

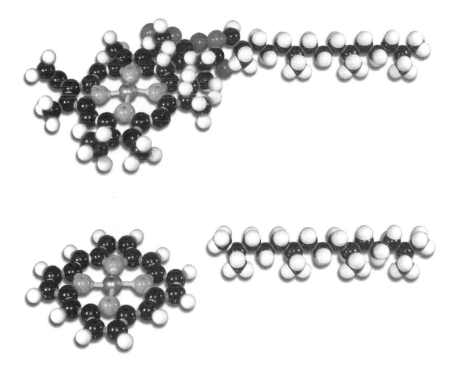

The top photograph shows a model of the chlorophyll molecule as produced by living plants. The one at the bottom shows two fragments from the breakdown of chlorophyll found in crude oil. Carbon atoms are shown in black, hydrogen in white, and oxygen and nitrogen in gray.

brave types have fought the public battles. My honor roll includes Martin Gardner, James Randi, Carl Sagan, and Cecilia Payne-Gaposchkin.

The first successful gas well in New York was drilled in 1821, thirty-five years before the more famous Drake oil well in northern Pennsylvania. Modern gas wells in the same region produce mostly from sandstone layers, from 2,000 to 5,000 feet deep. Wells are drilled rather quickly and (by petroleum industry standards) inexpensively by using a downhole hammer powered by compressed air. The hammer is similar to the noisy jackhammers used by construction contractors. If gas-bearing sands are discovered, liquids are pumped in under high pressure to open cracks radiating away from the well bore. The process is known as hydrofracturing, nicknamed "frac." Both New York and Pennsylvania wells are desirable locations for producing gas: close to major gas consumers, and the drilling costs are reasonable, making smaller wells economic. Another important help, both to gas producers and consumers, came in the 1980s when the courts declared that gas pipelines were common carriers. Previously, all natural gas was purchased at the wells by gas pipeline companies. The pipelines made a profit on buying and selling the gas as well as by supplying transportation. Now that gas pipelines are common carriers, operators of gas wells can sell gas directly to end users and the pipeline gets a fixed fee for moving the gas.

Fortunately, I had an advance warning about one of the hazards of oil and gas consulting. I was waiting in a lunch line at a petroleum meeting and two guys in front of me were discussing a failed drilling program. "We'll sue the geologist, too. He might have some malpractice insurance." I didn't carry malpractice insurance. My rationale was that someone might sue me for negligence, but if I gave a reasonable interpretation based on real observations it would be very hard to sue me simply for being wrong. During the early 1980s, I had the opportunity to advise a series of projects that drilled for natural gas in western Pennsylvania and New York. My formal letters of recommendation for drilling programs were long and they were boring, but they were not negligent.

The gas-reservoir sandstones in western New York and Pennsylvania were deposited parallel to ancient shorelines that ran northeast-southwest. A good drilling location was therefore northeast or southwest

An enlarged version of the construction-site compressed-air jackhammer, which can be used to drill wells. The hammer works at the bottom of the hole, pounding on a bit with tungsten carbide buttons. For wells less than five thousand feet deep, in relatively solid rock, and in the absence of large water-flow into the well, the downhole hammer is the least expensive drilling method. (Courtesy of Smith Bits)

of a better-than-average existing well. (A location adjacent to an existing well is called an offset.) Both the production statistics and the downhole wireline logs helped to identify the more profitable existing wells. In the town of Olean, the New York state government operates an excellent open library where well logs and initial production data can be photocopied. Selecting good drilling locations was more a matter of patience than skill.

Drilling routine gas wells is good bread-and-butter business, but some drilling programs throw in a chance to drill a barnburner—a big, profitable well. One of the programs I advised did hit a barnburner, the

largest gas well in the long history of New York State. Along the northern edge of a subsurface salt layer in New York, there is a fold, like a
crinkle at the edge of a carpet. The fold, and the oil and gas trapped in
the fold, was discovered by accident. It was a real accident: The drillers
were not expecting it and their drilling rig was lost in the ensuing fire.
One of my programs located a well farther along the crinkle fold. Our
driller knew about the problem but did not connect up the blowout
preventers: blowout, negligence, lawsuit. Fortunately, no fire. The well
was brought under control the next day. Many happy investors.

Unconventional Gas

There are several unconventional sources for natural gas. All of them
have been known for thirty years, some of them for centuries. They are
unconventional in the sense that rising gas prices have renewed interest in looking for commercial production. The first four gas sources in
the following section are useful but not likely to be large enough to res-
cue the North American natural gas market. The last item, gas hydrate,
could be large, but at the moment the world's only gas hydrate field is
producing only intermittently.

Swamp Gas

A variety of bacteria are capable of converting organic matter into
methane. The bacteria are not rare or exotic; all of us carry them in our
intestines. In the ground, bacteria generate methane from the grass
roots on down to a temperature where the bacteria do not survive. Before 1967, all bacteria were thought to die at the temperatures used in
pasteurizing milk, around 145°F. However, a few species of ther-
mophilic bacteria, capable of growth at higher temperatures, confused
the definition. At the moment, dirty-fingered ex-roustabouts suspect
that most or all methane-producing bacteria are the normal types that
die after a half hour at the Pasteur temperature. The unofficial name
"swamp gas" refers to the methane accumulations over swamps produced by these normal-temperature bacteria. Strange diffused light
flashes sometimes seen in swamps are probably from burning methane.
    In the shallow layers of sedimentary rocks, shallower than the top

of the oil window, bacterial gas can accumulate in commercially viable gas fields. The most-studied examples are found in rocks of Cretaceous age (around 100 million years old) in Montana and Alberta, although examples are known worldwide. None of the fields have been large enough to attract the attention of the major petroleum companies, but the shallow depth and associated low capital costs make biogenic gas recovery rewarding for smaller companies and Third World countries.

Coal Bed Gas

Many a miner has died from methane explosions in coal mines. The Davy safety lantern—and the canary—were used to detect both methane and low oxygen levels. As early as the 1930s, wells were drilled into gassy coal beds for commercial production of methane. Beginning in the 1980s, there was an expanded development of retrievable coal bed natural gas. In the United States, a temporary tax incentive helped initiate the program.

The coal itself is the source of the gas. In conventional oil exploration, burial of some types of organic matter (especially from marine algae) generates abundant oil. In contrast, plant material (dead leaves, tree trunks, lawn clippings) tends to generate only small amounts of methane and no oil. Coal bed methane is a significant source of natural gas only because the coal bed is a large accumulation of plant material.[4]

Coal bed gas, as of the year 2000, supplied about 8 percent of the U.S. total natural gas production. As so often happens with natural resources, a significant fraction of the total coal bed gas comes from a limited area. The San Juan Basin, in New Mexico and Colorado, has contributed 80 percent of the coal bed gas. A fairway, or sweet spot, measuring about ten miles by forty miles across the Colorado–New Mexico border currently produces 70 percent of the total U.S coal bed gas. However, coal bed gas production in the Powder River Basin in northeastern Wyoming is growing and may soon exceed that of the San Juan Basin.

In 1815, Humphrey Davy introduced a lantern for detecting explosive and toxic gases in coal mines; the oil-burning lamps are still being manufactured and used today. The leftmost two parts of the disassembled lamp on the bench are fine metal–mesh sleeves that surround the flame. Davy discovered that a metal mesh prevented the spread of an explosion away from the lamp flame. In "bad air" (usually carbon dioxide), the flame grew fainter. If natural gas (methane) was present, the flame burned higher and changed color. (Culver Pictures)

Basin-Center Gas

Normal exploration for oil and gas is a search for a trap, a place where oil or gas floats on top of the subsurface water but cannot migrate to the surface. In some areas, gas is so abundant that the porosity in an entire sheet of rock is gas filled and water is largely displaced. Whee! No more worrying about local oil and gas traps; just look for the thicker and more permeable streaks in the host rock. Although I did not appreciate it at the time, the 1980s drilling programs for natural gas that I advised in New York and Pennsylvania were conducted in a northern end of a gas-saturated layer that extended as far south as Tennessee. Other important basin-center gas accumulations exist in southwestern Wyoming and northwestern New Mexico.

The glory of having the pores in an entire layer filled with natural gas comes with a built-in disadvantage. There never seems to be enough gas or oil to saturate a big, thick, high-porosity, extensive reservoir rock. It is the thinner, less-porous rock layers that fill completely with gas. In particular, the ease of fluid flow is typically lower. In proper terms, these are low-permeability rocks; in oilfield slang they are tight sands. Hydrofacturing of the low-permeability reservoirs is almost invariably essential to the economic success of tight-sand gas wells. At one time, there was a U.S. price incentive to encourage tight-sand gas production. Today there is a more modest tax incentive for wells producing less than 90,000 cubic feet of gas per day.

The decision to drill—or not to drill—a gas well illustrates an interesting paradox. Put simply, it is a game of determining when a bird in the hand is worth two in the bush. Knowing that drilling an exploratory well for oil or gas is a risky business, try the following thought experiment. You are sitting at the head of a big conference table, with all the oil companies and independent investors attending. You have before you folders, each one describing an exploratory drilling opportunity during the year 2005, and you have sorted the folders with the ones likely to generate high rates of return on top. You put the best one out on the table and it is immediately snapped up. As you go down through the pile of folders, the bidding gets weaker. At some point, everyone around the table thinks the rest of the prospects are either dogs or turkeys and the auction is over. You have determined, by experiment, what rate of return is needed to attract capital into exploration wells. In current practice, that rate of return on invested capital is somewhere between 10 and 15 percent per year.

In its fully developed form, the investor is trying to estimate the present-day expected value of the discounted future cash flow. Let's take that apart:

- "Expected value" admits that some of these prospects are going to be dry holes, some might be small producers, and some might be large. Based on experience, the investor estimates

that one hundred prospects like this have generated seventy-
five dry holes, eighteen gas fields of a billion cubic feet, five
fields of ten billion, and two with a hundred billion cubic feet
of gas.

- "Cash flow" recognizes that not all the money is spent on day
  one; the investment may be spread over several years. Some of
  the return is many years in the future. Think of it as an imagi-
  nary bank account with a negative balance of borrowed money
  during the early years, a break-even or payback year in the
  middle, and a positive balance in the later years.

- "Discounted" uses an interest rate to convert payments in future years back to their present-day value.

In principle, the investor crunches the numbers and decides whether
to invest in this drilling opportunity. In practice, many experienced
people in the oil game do an approximate calculation in their heads in
a few seconds.

The point of this discussion is that basin-centered gas fields differ
from conventional gas fields in having a very high probability of success
and a low cash flow. As an investment, basin-centered gas wells look
more like federal bonds and less like biotechnology stocks.

Fractured Shales

The first U.S. commercial natural gas came from fractured shale
(mudstone) in upstate New York. Wells drilled in 1821 fed gaslights on
the streets of Fredonia thirty-eight years before Drake's first oil well.
Despite this long history, less than 2 percent of U.S. natural gas production comes from fractured shale.

As mentioned earlier, permeability, the ease of fluid flow, is lower in
fine-grained rocks. It's worse than it sounds: Cutting the size of the internal pores in half reduces the permeability by a factor of four. Reducing the pore size by a factor of ten lowers the permeability by a factor of
one hundred. In general, the permeability depends on the square of the
grain size. Compare a sandstone with one-millimeter grains to a shale
with one-micron grains. While the size ratio is a thousand, the permeability difference is a thousand squared, or a factor of one million higher

This is open-flame natural gas street lighting in Findlay, Ohio, in 1885. In the same year, the more efficient gas-mantle light was invented in Europe. (Culver Pictures)

in the sandstone than in the shale. Gas flow through the shale itself is too slow to be commercially valuable. However, if the shale comes equipped with an extensive set of natural fractures, the fractures contain gas and can conduct gas that is released slowly from the bulk shale.

There are oil equivalents to fractured-shale gas reservoirs. The famous (infamous, actually, for not being very profitable) Spraberry oilfield in West Texas covers three counties. The oil is produced through natural fractures from siltstone. As the name suggests, siltstone has a grain size intermediate between sandstone and mudstone. Spraberry generated a large number of just barely profitable wells.

On the downward side of Hubbert's peak, every little bit helps, including fracture-shale gas. However, fractured shales are not a candidate for solving a major portion of our energy needs.

Gas Hydrates

These are something bigger than "every little bit." Until now, only a trivial amount of gas has been produced from gas hydrates. Further, there is no agreed-upon methodology for extracting gas from hydrates.

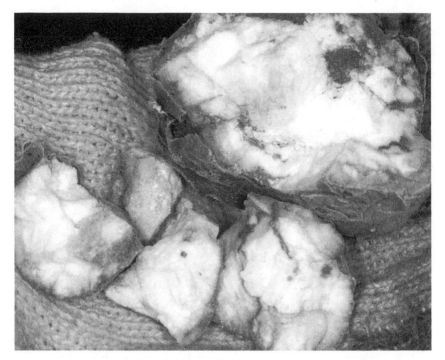

Natural gas hydrates, brought up from the seafloor off Oregon. The white
material is the hydrate and the dark inclusions are embedded sediment. The
background is a knitted mitten.

Some experts estimate the gas in hydrates to be about a factor of one
hundred larger than all the world's conventional gas reserves, although
recently some doubts have arisen.[5]

Gas hydrates, also known as gas clathrates, are part of an extensive
family of crystalline solids that results from the freezing of a gas-water
mixture. The dominant molecule in natural gas, methane, is only one
of a dozen molecules known to form clathrates with water. Others include nitrogen, oxygen, hydrogen sulfide, carbon dioxide, chlorine,
chloroform, and even inert gases like argon and xenon. The methane
hydrate contains four methane molecules and twenty-three water molecules. As an aside, methane hydrate fits the official definition of a
mineral: a naturally occurring substance with a definite crystal structure. Some reader of this book might want to go through the trouble of
getting methane hydrate a proper mineral name like "kvenvoldenite."

The first clathrate discovered was chlorine, in 1810. During the 1930s and 1940s, methane hydrates were identified as a nuisance that blocked natural gas pipelines. Around 1970, news began to circulate about extensive gas hydrate deposits under the arctic permafrost and under the outer continental shelves. My earliest education on the topic began in a four-way conversation at a party: Another Princeton faculty member, Dick Holland, and I were loudly exchanging what little we knew about the rumors, knowing that the other two listening, with big grins, were Hollis Hedberg (Gulf Oil) and Creighton Burke (Mobil Oil). Eventually, Hedberg and Burke couldn't stand it and started telling us the real story.

Subsurface methane hydrates are important in ways other than as an energy resource. Methane hydrate is not stable at room pressure or at room temperature, so that a lowered sea level or increased temperature would release the methane gas. Some subsea landslides have been attributed to methane release. Even more important, methane is a powerful greenhouse gas, much stronger than carbon dioxide. Any geologic event that lowered sea level or raised temperature on extensive methane hydrate deposits could cause severe global warming.

There are two somewhat different settings that contain extensive methane hydrate deposits.

- At high latitudes on land, sediments with frozen water—permafrost—extend from the surface down to depths of 600 to 3,000 feet. Beneath the permafrost there is often a layer of methane hydrate. Almost automatically this limits recovery to the northern hemisphere because the only suitable land in the southern hemisphere is Antarctica, and we are currently not considering recovering resources there.
- The outer parts of the continental shelf at water depths from 2,000 to 8,000 feet also contain methane tied up in subsea hydrates. The total methane in this setting could be a factor of ten larger than the continental deposits.

Subsea methane hydrate deposits extend from the poles all the way to the equator for an interesting reason. Most of the volume of the present-day ocean is filled from the bottom up with water made heavy by a slightly higher salt content and chilled to within a few degrees of freezing. (A curious by-product: When I joined the oceanography department at Oregon State, the first thing I was told was: Don't go swimming! Earlier, a professor of English arrived a few days before the fall semester and went down to the coast and dove in. He never came up. Summer winds strip away the warm surface water, exposing water as cold as 39°F.) On the outer continental shelf off Oregon there is a spectacular site where the methane hydrates are exposed on the sea floor, a place now named "Hydrate Ridge." Lumps of methane hydrate and sediment impregnated with hydrate are observed floating to the surface. A recent detailed study of Hydrate Ridge with reflection seismic data and nine drilled boreholes indicated that the gas hydrate zones are thinner and more discontinuous than had been expected.[6]

To date, the only commercial methane production from hydrates has come intermittently from the Messoyakha field in north central Siberia. The field began producing in 1969. The initial production was from a conventional natural gas reservoir immediately beneath a gas hydrate layer. Because the gas pressure in the reservoir did not drop as fast as expected during gas production, it became clear that gas was being contributed by the breakdown of methane hydrate. However, despite several attempts to enhance gas production, Messoyakha has not become a major producer. Since 1980, gas production from the entire field has been less than a million cubic feet per day. For comparison, a million cubic feet per day is typical production from a single midsize Texas gas well.

Because of the apparent worldwide extent of gas hydrates in the ground, thought has been put into recovery methods. Methane hydrate formation depends on low temperatures, high pressures, and water. The obvious ways of breaking down the hydrates to recover the gas are raising the temperature, lowering the pressure, or "stealing" the water. Each option has pluses and minuses.

Warming: The first worry was that breaking down the hydrate by heating it might require more heat that you would get later by burning the methane gas. Not to worry; heating requires about 7 percent of the energy content released by burning the gas. An interesting alternative is the use of geothermal heat. Virtually anywhere, including the arctic, deep wells in sedimentary rocks will produce hot water. The geothermal generation of electricity requires near-boiling water, but in recovering methane hydrates, we need a temperature rise of only 10°F. In short, low-grade geothermal heat will do. Finding a heat source is easy, however, compared to distributing the heat into the gas hydrate layer. The normal pore space within the sediments is plugged up by the gas hydrates, so simple injection of a hot fluid into the hydrate layer probably will not work.

Pressure drop: If there is a conventional gas reservoir immediately beneath the gas hydrate layer, then producing the conventional gas will initiate gas hydrate breakdown. This seems to be the major production mechanism at Messoyakha. However, once the reservoir pressure is low enough to decompose methane hydrate, there is only a modest pressure gradient left to move gas to the production wells. Vacuum pumps have occasionally been installed on nearly depleted conventional gas wells. However, we normally count on thousands of pounds per square inch to move gas around in a reservoir; even a perfect vacuum pump can add only another 14.7 pounds per square inch.

"Stealing" the water: Give the hydrate a chemical choice that will attract water even more strongly than methane. In a broad sense, we are looking for "antifreeze," although most of the candidates are not suitable for use in automobile radiators. The first experience came in removing hydrates that clogged natural gas pipelines. Methanol (wood alcohol, a toxic liquid) became the popular choice for cleaning pipelines. Ten attempts to stimulate wells at Messoyakha by injecting thousand-gallon slugs of methanol did not significantly increase gas production. Another "antifreeze" is a brine made from any highly water-soluble compound. One attempt at Messoyakha used concentrated calcium chloride solution along with methanol, but the results were no better than with straight methanol. Another possible natural help: Some deep oilfield brines are highly concentrated solutions of sodium chloride and calcium chloride; they might be used to decompose gas hydrates.

Gas hydrates today are an opportunity for some young person to become richer than Bill Gates. The methane hydrates already identified are larger than all the world's oil, gas, and coal combined, although the recent detailed study at Hydrate Ridge raises the possibility that the world gas hydrate estimates may be too optimistic. A company built around a patented and successful hydrate extraction process will carve your name over the entries of museums, libraries, football stadiums— whatever you like. It won't be easy; the major oil companies have known about the gas hydrates since 1970 and so far they haven't announced a promising extraction strategy.

I'm not working on the gas hydrate problem, but if I were, I would start by taking a hint from nature. Some of the subsea gas hydrate areas seem to be spontaneously bubbling gas. Even if they are not leaking gas at commercially valuable rates, it's a hopeful start. I'd try to diagnose what's driving the natural leaks and then see whether I could then harness the process.

The contributions of swamp gas, coal bed gas, basin center gas, and gas from fractured shales are modest, but welcome, contributions to our energy supply. They are modest on the scale of the North American gas supply but potentially highly profitable to small companies or to individuals. Gas hydrate deposits are potentially large, but as yet we have no engineering successes for gas hydrate production.

Water, Water, Everywhere

In many areas, a small amount of water is produced from several types of gas reservoirs. The water ranges from being a nuisance to a major problem.

- Just as water vapor appears as humidity in the atmosphere, water vapor also shows up in natural gas. Pipeline operators have restrictions on the amount of water vapor they will accept. One reason is that at low temperatures the water vapor and the natural gas combine to plug the pipeline with gas hydrates. That problem is most severe in the winter. However, gas producers complain that pipeline company inspectors generally show up in midsummer when the pipeline doesn't want to accept much gas. There are commercial water-removal systems that absorb the water into antifreeze liquids, but they are complex and expensive. A simple and inexpensive system for dehydrating a million cubic feet of gas per day would be most welcome.

- Liquid water, often as salt brine, frequently comes into the producing wells. If the water accumulates in the well bore, eventually the water column exerts enough pressure to stop the gas flow. Various tubing arrangements, soap sticks, and other ingenious methods go into getting the water to the surface. Once at the surface, salt brines are usually considered hazardous wastes. Gas well operators often pay several dollars per barrel to have the salt water trucked to and disposed of at a licensed waste-disposal site. (For some years, I have been telling my students to consider acquiring depleted natural gas reservoirs and getting them licensed for hazardous waste disposal.) Local salt-water disposal wells are common in oil fields, but gas wells usually do not produce enough water to justify drilling dedicated disposal wells.

- Coal bed natural gas has a unique water problem. In most areas, the fractures in the coal beds are initially filled with water. If that initial water is pumped out, methane gas starts to bleed out of the coal and travel through the fractures to the producing wells. There is initially a large amount of water, and water production continues even as economic amounts of gas are recovered. The water can either be fresh water or salt water. Drinking water, for people or for animals, contains less than five hundred parts per million of salt. Water up to one thousand parts per million can be used to irrigate crops. Above that level, a disposal system is needed for the water. In particular, the Powder River Basin in Wyoming contains some enormously thick coal beds that are attractive for producing coal bed methane. In that semiarid climate, usable water is economically welcome. Sagebrush flats become hay meadows. However, some of the coal bed water in the Powder River Basin is salty. In that same semiarid climate, evaporation ponds will slowly get rid of the water, but there are environmental objections to open-air saline ponds.

Transporting the Gas

The oil market is global. Supertankers transport oil halfway around the world for two dollars per barrel. In contrast, natural gas is largely a continent-by-continent market. Natural gas pipelines move huge amounts of gas on land, but gas transport across oceans is limited.

The only practical method for moving natural gas on ships is to cool it to −259°F to convert it to liquid natural gas, sometimes called LNG. In the United States, LNG got off to a bad start. Any new industry or technology is severely damaged by a spectacular accident: The Hindenburg disaster put the kibosh on passenger transport by rigid airships. In Cleveland in 1944, liquid natural gas leaked into the sewer system and exploded; 128 fatalities resulted. On Staten Island in 1973, an explosion during repairs to a LNG facility killed thirty-seven workers. Some revisionists today claim that the Staten Island fatalities were a "construction accident," but I remember reading about it at the time and I considered it an LNG accident. In January 2004, an explosion at a gas-liquefaction facility in Algeria killed between twenty and thirty people. Again, the explosion may have been in a steam boiler that powered the gas refrigeration and not in the LNG itself, but some doubt remains. Today, there are four U.S. seaports receiving LNG: Everett, Massachusetts; Cove Point, Maryland; Elliott Island, Georgia; and Lake Charles, Louisiana.

LNG imports amount to about 5 percent of U.S. natural gas imports; most imports come by pipeline from Canada. Although Canada, the United States, and Mexico all are major gas producers, Canada exports to the United States and the United States exports gas to Mexico. NAFTA requires all three countries to continue their historic pattern of sharing natural gas. During a serious gas crunch, however, NAFTA might get wadded up and pitched in the wastebasket.

As mentioned earlier in this chapter, the Hubbert data and methods that work reasonably well for oil are not readily applied to natural gas. The best study of gas supply is an analysis by Matthew Simmons, who heads a large energy banking firm in Houston.[7] He obtained detailed production data from the Texas Railroad Commission for gas wells in fifty-three Texas counties. Simmons's conclusion is that we

A liquefied natural gas (LNG) tanker ship leaves Boston Harbor in January 2004 after unloading at the Everett LNG terminal. (Michael Dwyer, Associated Press)

have been keeping levels of U.S. natural gas production almost constant only by drilling an ever-increasing number of wells each year. Several other observers have reached the same conclusion; it is called the Red Queen strategy[8] because we are running as fast as we can to stay in one place.

Simmons also points out that newer gas wells have shorter and shorter productive lives. In part this results from better well-completion techniques, especially improved hydrofrac treatments. If future earnings are discounted by an interest rate, as explained earlier in this chapter, then the well operator has a motivation to make the well pay back its capital cost as rapidly as possible. I keep asking people in the industry whether today's lower interest rates would cause wells to be drilled farther apart and have longer lifetimes. My usual reward is a blank look (but not from Simmons), implying that I don't understand the problem.

As discussed in Chapter 2, prices are expected to become irregular and chaotic when the demand approaches the system's capacity. The figure on page 32 shows that U.S. natural gas prices were reasonably stable through 1985 and have undergone increasingly wide swings ever

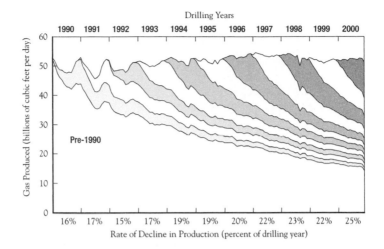

Recently drilled U.S. natural gas wells show faster rates of decline than older wells. In part, this could result from more aggressive use of hydrofracturing techniques. However, the full explanation for the increasingly rapid decline is not obvious. (Jeffrey L. Ward, based on data supplied by Petroleum Information Corporation; original chart prepared by EOG Resources, Inc., copyright © 2002)

since. Natural gas price excursions were particularly wild during the unusually cold winter of 2002–03—prices doubled over one weekend.

Major gas consumers can lock in a gas price if there is a producer willing to sign a long-term contract. Anyone can purchase a contract in the commodity futures market to assure a definite price on a future date for either buying or selling natural gas. Those of us who heat a house and run a cookstove usually don't go to the trouble of locking in a gas price. Thirty years ago, natural gas was treated as a regulated pub-lic utility. The current political enthusiasm for deregulation leaves many small consumers exposed to increasingly wild market swings. Af-ter the California and Enron disasters, some people yearned for the good old days of regulated utilities. My only advice is to be careful. Much of the chaotic pricing is due to demand approaching capacity. Returning natural gas to a regulated utility will not legislate away the demand-capacity problem.

I should confess that my 2001 book *Hubbert's Peak* was overly opti-mistic about the natural gas supply. Around 1970, price increases changed natural gas from a waste by-product to a valuable resource. Exploration

expanded into areas beyond the oil window, areas that are now—
or once were—hot enough to convert crude oil into natural gas. Developing the full oil potential of the United States took about seventy
years. My mistake was presuming that the natural gas potential would
take a comparable length of time. Instead, between 1980 and 2002, the
best of the natural gas targets were drilled. We're now being served
leftovers.

Five

Consider Coal

Coal is the best of fuels; it is the worst of fuels. Best, because it is much less expensive per unit of energy. Worst, for a long list of reasons: killer smog, acid rain, atmospheric carbon dioxide, mercury pollution, acid mine drainage, and a choice between hazardous underground mines and surface-disturbing open-pit mines. In what passes as dark humor among energy providers, a nuclear engineer once wrote a fake safety memo claiming that a coal-fired electric generating plant would be unacceptably hazardous. Despite coal's drawbacks, there are six thousand active coal mines in the United States. Worldwide coal reserves are large enough to continue present rates of production for a few hundred years.

The Industrial Revolution was fired with coal. Coal served both for heating and for the chemical conversion of iron ore into iron and steel. Discovery of new coalfields was of major importance to industrializing England. The coal beds were formed only during certain episodes in geologic history. Many of the geologic periods that we use today were originally defined during the search for British coal.[1] The United States and the former Soviet Union had, and still have, the world's largest coal reserves. They became the two biggest industrial economies in large part because of easy access to large quantities of coal.

The crowning symbol of the Industrial Revolution was the blast furnace. Iron ore, pretreated coal, and limestone were fed into the top of the blast furnace, and preheated air was blown into the base. Liquid iron and a glassy slag poured out the bottom of the furnace. Coal for the blast furnaces was pretreated in large ovens to produce charred lumps known as coke. The gaseous and liquid by-products from the coke ovens were a witches' brew, but whole industries were founded to utilize the coke by-products. The gases would burn and were used for lighting or heating. Ammonia was converted into fertilizer. Liquid hydrocarbons, called coal oil, were used primarily for kerosene lanterns. (My father's generation continued to refer to kerosene as coal oil for fifty years after petroleum had become the dominant source of kerosene.) The most spectacular industrial innovation, a family of new synthetic dyes, used the heavy tars produced by the coke ovens. The aniline dyes generated a range of fabric colors not previously available from natural products.

Once the coke-oven by-products had shown the way, new chemical products came in profusion. Coal was reacted in different ways to generate useful products. Burning coal in the presence of water vapor generated mixtures of carbon monoxide and hydrogen variously known as town gas, producer gas, water gas, coal gas, or synthesis gas. The carbon monoxide gas is highly toxic and these coal-derived gases are no longer distributed in cities. However, the reaction has become an initial step in both existing and potential future processes.

For household use today, coal-derived gases have been replaced by methane from natural gas wells. There is a process for making methane from coal known as synthetic natural gas (the product is called SNG because it is difficult to say "synthetic natural gas" and keep a straight face). Development of SNG is not on our current agenda, but it may appear as natural gas becomes scarce.

The mixture of carbon monoxide and hydrogen can be used, without further modification, as a chemical reactant. For instance, blast furnaces have largely been replaced by a process called direct reduction of iron ores. It works thisaway: The first reactor converts coal, air, and water vapor into carbon monoxide and hydrogen, and a second reactor

Blast furnaces produced iron by feeding in iron ore, coke (processed coal), and limestone in the top and injecting a blast of preheated air near the bottom. A pool of liquid iron formed at the bottom and could flow out for further processing. This blast furnace was near Pittsburgh; the photograph was probably made before 1960. Although most blast furnaces have been replaced with newer technologies, several are still operating around the world. (Culver Pictures)

When heated, some grades of coal form a porous char known as coke, an essential ingredient for blast furnaces. In this photograph, glowing hot coke is being pushed out of an oven into a railcar. (Culver Pictures)

uses those gases to reduce iron ore to iron. (The process ought to be called indirect reduction. The blast furnace is actually more direct, but we'll ignore that.) Other metals can also be recovered from their ores by direct reduction.

A process of some historical importance converts the gases from coal reactions into high-grade gasoline. The Fischer-Tropsch reaction uses an iron-rich catalyst to convert gases into liquid hydrocarbons. During World War II, Nazi Germany had almost no access to crude oil. Gasification of low-grade coal and the Fischer-Tropsch conversion produced 100-octane aviation gasoline as well as diesel fuel. An upgraded version of the same process was used in South Africa during the apartheid era. Although the process has a politically incorrect ancestry, all of us atop Hubbert's peak may have to turn to coal-derived gasoline and diesel fuel to fill part of the gap caused by dwindling world crude oil production.

Reacting coal with steam and a limited amount of air produced "coal gas," a dangerous mixture of carbon monoxide, hydrogen, and methane. This 1870s engraving shows coal gas being delivered for household lighting in Paris (France, not Texas). (Science Photo Library)

Right now, the fastest-growing coal gasification process is one optimized to produce hydrogen gas. There are two substantial existing markets for hydrogen:

- Petroleum refining uses hydrogen to upgrade the larger and heavier molecules in crude oil to the smaller hydrocarbons in gasoline.
- Hydrogen is reacted with atmospheric nitrogen, using a nickel catalyst, to form ammonia, which is then converted to fertilizer and other nitrogen-containing products.

During the apartheid era, South Africa produced gasoline from coal. This photograph of the Sasol plant was made in 1975. Although South Africa now has access to imported oil, this plant today turns coal into petrochemical feedstocks. (© Gerald Cubitt)

Previously, the cheapest means of producing hydrogen was by breaking down methane from natural gas. However, in some markets the demand for natural gas has exceeded the production capacity. Texaco, now part of ChevronTexaco, optimized the older coal gasification processes to produce hydrogen; the first plant was licensed in 1954. In China, there is an abundance of coal, only a modest supply of natural gas, and a huge need for fertilizer. As of 2001, Texaco had sold eight coal gasification plants to China, and more are on order. Inside the Texaco process, coal reacts with pure oxygen and steam at a very high temperature, and then the resulting gases are cooled and hydrogen is separated. The by-product is carbon dioxide, which the Chinese—to the frustration of environmentalists—are currently sending up the smokestack. The largest market for carbon dioxide is for enhanced recovery of oil from existing oilfields. Selling the carbon dioxide would improve the economics. Next best would be injecting it into a natural underground reservoir instead of releasing it into the atmosphere.

On the horizon is the production of dimethyl ether from coal. (The ether used as an anesthetic is d*iethyl* ether.) Currently, the leading use of dimethyl ether is as a propellant in cans of hair spray. However, dimethyl ether is an almost ideal diesel fuel. Producing dimethyl ether from coal begins with the same Texaco process for producing hydrogen, but the hydrogen and carbon monoxide are converted to dimethyl ether. Current cost estimates to produce diesel fuel in this way are between $1 and $2 per gallon, before the taxes are added. A 14,000-barrel-per day pilot plant is planned in Ningxia, China.[2]

Despite the variety of useful products from coal, most coal is simply ground up and burned as a source of heat. About two-thirds of all coal mined in the United States is burned to generate steam, which turns turbines to generate electricity. Because coal is a relatively inert solid, cleaning the coal before burning to remove sulfur or mercury is almost impossible. The exhaust gas in the smokestack is chemically more accessible, but scrubbing a large amount of gas, to reduce pollution by carbon dioxide, sulfur, and mercury, is not an easy task. It is sometimes cheaper to transport low-sulfur coal from Wyoming than to remove the sulfur from Appalachian coal.

The first coal-gasification plant utilizing the Texaco design was used for generating electricity, with reduced atmospheric pollution as the objective. The plant was operated by Southern California Edison for five years beginning in 1984. It was an engineering success and an economic failure. After five years, the facility began a second, less-glamorous career incinerating sewage sludge.

The Origin of Coal

During growth, green-leaved plants convert atmospheric carbon dioxide into organic material plus oxygen gas. Bacteria then oxidize discarded leaves, and eventually the entire plant, back into carbon dioxide. When the two processes balance, there is no net storage of carbon and hence no contribution to a potential coal deposit.

In bogs or swamps, plant growth exceeds decay and an organic layer accumulates: a possible future coal deposit. However, an organic-rich bog is a chemically unusual place. As the underlying organic layer gets thicker, plants can no longer obtain mineral nutrients from the soil. The bog or swamp becomes a low-nutrient environment. Plants extract the tiny flow of plant nutrients that comes dissolved in the rainwater. One biochemist at the University of Minnesota found that water flowing out some of the ten thousand Minnesota swamps was cleaner than the distilled water in the chemistry building. Windblown dust and volcanic ash also contribute, but life is tough in the swamp. Some plants, for instance pitcher plants, trap and digest insects to supplement their nutrient supply.

When Alfred Wegener wrote his first discussion of continental drift in 1912, he used coal beds as an indicator of ancient tropical climates.[3] Like most of us, he associated lush plant growth with tropical jungles. But those bags of peat moss in the garden store come from Canada, not Brazil. Peat accumulates in bogs, even on gently sloping hills in Scotland. Blocks of peat can be cut, dried, and burned as fuel. However, the greatest fame of Scottish peat comes from Scotch whisky: Sprouted barley is dried over peat fires, and the peat smoke winds up as a recognizable flavor in the whisky. As a veteran of five sessions of teaching field geology during spring breaks in Scotland, I can absolutely certify that the climate isn't tropical. For starters, we have to admit the possibility that some coal might have been formed in cold climates.

Measurements of the inclined magnetic field in ancient sedimentary rocks allow us to determine ancient latitudes. As it turns out, most coals are from latitudes, north or south, greater than 40°, in sharp contrast to Alfred Wegener's hypothesis. In particular, the hundred-foot-thick coal beds in the Powder River Basin in Montana and Wyoming were deposited when North America briefly reached 58° north.[4] "Briefly" in this context means ten million years.

Once formed, organic-rich bog and swamp layers turn into coal only if they are buried. The fate of most surface materials is to erode

Peat in Ireland, 1984, being cut in the trench on the left and stacked for drying. Peat does not necessarily come from flat swampy bogs. (© Farrell Grehan/Photo Researchers)

away; soils and lake sediments are underrepresented in the geologic record. An effective aid for burying coastal swamps would be a rapid rise in sea level, which leads to an interesting detective story. When the geologic time scale was being developed in England during the nineteenth century, geologists gave the name Carboniferous to the period when the major British coal beds were deposited, roughly 280 to 310 million years ago. Coal beds in the eastern United States turned out to be of the same age. This came as surprise then but is no surprise today: During the Carboniferous period, Europe and North America were joined at the hip. (In American usage, the coal-rich Carboniferous beds are usually called Pennsylvanian.)

During the late nineteenth and early twentieth centuries, careful examination of Carboniferous rocks showed a sequence of layers that repeated several dozen times. Every time the sequence repeated itself, there was a coal bed. Instead of naming each coal bed, they were simply numbered; the most productive coal bed near Pittsburgh is the

Number 6 Coal. The repetitive sequence contained a major clue to the coal's origin: The layers immediately above the coal typically contained shells of marine fossils. The absence of marine fossils below the coal hinted at deposition above sea level. Aha! The coal probably was a coastal bog or swamp deposit buried either by the rise of sea level or the dropping of the land surface. But why?

This next secret mustn't get out: We scientists cover our ignorance by pasting an impressive-looking name on something we collectively don't understand. In 1930, the name "cyclothem" was coined for geologic units in the repeating coal-bearing sequence. The Carboniferous rocks contained cyclothems, but naming them did not explain them. When I was a student, we heard learned debates about what, if anything, cyclothems actually were. Either the land surface twitched up and down or there were major oscillations in sea level. Maybe both.

In 1961, I had the thrill of hearing a talk by the geologist who finally cracked the problem. Harold Wanless, of the University of Illinois, wondered whether sea-level changes associated with the waxing and waning of continental glaciations could have generated the cyclothems. He knew that major glacial episodes during the last million years had repeatedly raised and lowered sea level, and he wondered whether something similar could have happened during the Carboniferous period 300 million years ago. The southern continents—Africa, South America, Australia, Antarctica, and India, which were then the Gondwana supercontinent—had evidence of extensive and repetitive glaciation. However, the Gondwana glacial deposits were reported to be of Permian age, not Carboniferous. The Permian period is the one immediately after the Carboniferous, but we have to be careful. Those "periods" are twenty and thirty million years long. The plant material that was deposited in the coal swamps can't be explained by glacial episodes that happened twenty million years later.

Wanless arranged to go to Australia, where the sequence of southern hemisphere glacial deposits was well developed. Although the southern hemisphere glaciation had long been assigned to the Permian period, the Australian geologists explained that the age assignment was actually in doubt. There were early Carboniferous fossils in the layers *beneath* the glacial gravels, but there were no visible fossils within the actual glacial deposits. Part or all of the southern hemisphere glaciation could have happened during the late Carboniferous.

The Carboniferous glacial sequence in Australia contains fifty beds of glacial debris (mostly gravel) separated by layers of finer-grained sediment. Each of the fifty gravel layers presumably results from an advance and retreat of continental glaciers. Building a major continental ice sheet has a global consequence: Enough water is stored in the glacial ice to lower sea level. (As an example, if the present-day ice sheets covering Greenland and Antarctica were to melt, world sea level would rise about two hundred feet.) Each time the southern hemisphere glaciers retreated, coastal swamps in Europe and North America were buried by the global rise of sea level.

The last piece of evidence that closed out the cyclothem detective story was the number of glacial cycles: roughly fifty glacial gravel layers in Australia and fifty-one cyclothems in North America. The counting isn't exact; there are some partial cycles in the record. However, the agreement in age and in the number of glacial cycles essentially ended the long debate: The major coal beds of Europe and the eastern United States are coastal swamps buried by sea-level changes caused by the southern hemisphere glaciation.

Ever since hearing Wanless in 1961, I have pursued a silly little hobby. Those sea-level oscillations he identified have to be global. Any beds of late Carboniferous age, whether they contain coal or not, are going to be influenced by sea-level oscillations. So, whenever I see a mountainside that looks as if it has stripes, where each stripe is a hundred or two hundred feet thick, I wonder whether it is of late Carboniferous age. The later Carboniferous sedimentary rocks, even from miles away, look distinctively more striped than rocks of any other age.

My hobby seemed a little less silly after 1983. Bob Berner, Tony Lasaga, and Bob Garrels published a model (known as BLaG) of the earth's atmosphere and ocean chemistry through time.[5] An unexpected result from their model was that the atmospheric oxygen *doubled* during the Carboniferous. Normally, the bacteria that oxidize organic matter reverse plant photosynthesis and there is no net effect on atmospheric oxygen or carbon dioxide. However, if carbon-rich plant tissues escape bacterial decay and are buried to form coal, oxygen from photosynthesis is stranded in the atmosphere. A route for oxygen to get out of the atmosphere is through the chemical alteration in soils of reduced (typically green or black, called ferrous) iron-rich source minerals and to produce sediments containing oxidized (usually red, ferric) iron. Sedimentary rocks from the two periods following the Carboniferous, the Permian and the Triassic, are typically red, as eloquently stated by John McPhee.[6]

Although initially the link between Carboniferous coal and Permo-Triassic redbeds sounded like one of Rudyard Kipling's "Just So" stories, Bob Berner made a serious attempt to get the oxygen doubling out of the BLaG model. Berner announced that his attempt failed; in his opinion oxygen pressure twice as high as today's oxygen is a robust property of the model.[7] This "failure" is an interesting insight into the workings of science. Most scientists keep an informal mental list of ideas arranged from the most probable (unlikely to be overturned soon) to ideas that might possibly be correct but are at risk of being overturned. When I learn that a capable scientist spent a couple of years in a serious but unsuccessful effort to disprove an idea, I move the idea up a few notches in my mental probability list. I now hold the opinion that depositing the Carboniferous coal beds caused the later deposition of red sediments.

Despite the early economic importance of the late Carboniferous coal, there are coal deposits of other ages. The thick coal beds in Wyoming and Montana are much younger. Elsewhere there exist much older coal beds. When I was a graduate student at Princeton, Professor A. F. Buddington mentioned seeing a coal of Precambrian age in the Soviet Union, dating from the time before there were fossils with hard shells. In 1957, most of us thought that there was almost nothing living during the Precambrian. Questions flew from the audience, essentially halting Buddington's lecture. He disappeared into his office and returned with a chunk of shiny black coal. About half of the coal had been converted to crystalline graphite. As a consequence, the Precambrian coal would just barely burn, but the Russians were mining it anyway, raising a lesson we would do well to remember: Difficult times encourage desperate recovery measures.

Two things contribute to the diversity of coal deposits: the starting plant material and the burial temperature that "cooks" the soft peat into solid coal. Coal petrologists use a huge vocabulary, which I don't understand, to describe this diversity. This is sort of poetic justice;

we noncoal geologists use a specialized vocabulary that fills a 740-page dictionary.

The later part of geologic time, the time that does have hard-shelled fossils, is divided into Paleozoic, Mesozoic, and Cenozoic. Those few scholars who study ancient plants have to keep reminding us that the "zo" in those words refers to animals. If you want to study fossil plants, there are Paleophytic, Mesophytic, and Cenophytic eras, with time boundaries different from the zoo.[8] During the middle of the Paleozoic on the animal clock, plants—largely reproducing from spores—first colonized the land: This initiated the Paleophytic era. The Mesophytic runs from mid-Permian to mid-Cretaceous and is dominated by ferns, cycads, and conifers. Angiosperms, the modern flowering plants, developed during the Cretaceous, and that Cenophytic flora still dominates today.

In addition to changing over time, plants contain a lot of different parts that can wind up in different places in the swamp. Wood, leaves, moss, grass, subsurface rhizomes, algae, and even spores generate different varieties of coal. Cannel coal consists mainly of spores and pollen. Cannel coal is mined and is a favorite for burning in home fireplaces. I even have a box of cannel coal in my garage, left over from my Y2K emergency preparations.

One coal oddity I can't pass up: jet. Jet is a hard, smooth, deep black coal that makes up small lenses in sedimentary rocks near the Scottish-English border. Especially during Victorian times, jet was cut into jewelry worn by women in black clothes mourning the death of a husband. Queen Victoria wore mourning for years after Prince Albert died; cynics claimed that she looked good in black. Jet jewelry is no longer in fashion, I once searched through an enormous flea market in Glasgow without finding any jet jewelry. Some is offered on eBay, but you have to be careful. Jet passed its name on to a color: jet black. Some of the eBay offerings are described as "jet black," which does not necessarily mean that they are jet.

In addition to differences in starting materials, coal beds are further diversified by their burial histories. After burial, and especially with the temperature of maximum burial, coal beds pass through lignite, sub-bituminous, bituminous, and anthracite coal. As it moves along the sequence, coal becomes increasingly carbon rich. Anthracite is more than 90 percent carbon. What we use commercially is mostly bituminous coal. In 1915, David White pointed out that coal and petroleum require about the same burial temperatures.[9] The temperature at the center of the oil window corresponds to the temperature of midrank bituminous coal. Anthracite coal is formed below the oil window, where the only hydrocarbon to be found is pure methane.

One of coal's serious nuisances is the presence of sulfur. Sulfur occurs in coal both in organic compounds and as fine-grained crystals of the inorganic mineral pyrite, FeS_2. In liquid crude oil or in natural gas, it is possible to remove the sulfur before marketing the product. In solid coal, no such luck. Part of the motivation for mining coal in Wyoming is a sulfur content lower than the eastern U.S. coal beds.

Originally, coal-burning facilities simply let the sulfur go up the smokestack. When I moved to Princeton in 1967, I could often smell (or taste) sulfur dioxide from the coal burned in Trenton. Sulfur dioxide released in the air eventually gets further oxidized to sulfuric acid and becomes a major contributor to acid rain. I had a real surprise when I looked back at Clarke's (1924) "Data of Geochemistry."[10] Large rivers in the northern hemisphere had what I expected: bicarbonate first, sulfate second, and chloride third. But, surprise, surprise, many southern hemisphere rivers contained almost no sulfate. Virtually all that northern hemisphere sulfate was industrial, even before 1924. Especially in the last twenty years, the switch to low-sulfur fuels has remedied part of the problem. I no longer smell (or taste) Trenton.

Pyrite raises another problem: acid mine drainage. Not all the pyrite gets hauled away with the coal. Some noncommercial coal and some associated mudstones are left underground or piled in heaps at the surface. Gradually, with the help of bacteria, the pyrite weathers and oxygen from the air converts the pyrite sulfur into dilute sulfuric acid. Streams contaminated with acid mine drainage can become acidic enough to kill fish. Robert Kleinmann, when he was one of our graduate students at Princeton, found that the acid-tolerant bacteria were easily killed by small amounts of common detergents.[11] One remedy that helps somewhat is to go around after every rain and spray detergent over the waste heaps. Underground, it isn't as simple.

Coal is mined both in underground mines and in open pits. There is a long list of ways to die in an underground coal mine; I'll spare you the gory details. Only once have I been underground in a coal mine and once was quite enough, thank you. I was on a student trip from the Colorado School of Mines; the mine was at Ludlow, in southern Colorado. At the surface, I noticed that all the miners seemed to be short and Hispanic. Underground, it became clear that Hispanic didn't matter, it was short that made the difference. Where the coal seam had been mined out, the space was large enough for me to stand up straight (5' 11"), but the wooden support timbers filled up about six inches of that space. Every once in a while I would forget and bang my head on a timber. Even with a hard hat on, it soon gets old. The worst part: After you get out of the mine, you blow your nose and notice a big black area of coal dust on your handkerchief. Because of the large number of hazards, there is no way to make a coal mine into a genuinely safe workplace. The most effective response is reducing the number of miners underground; some modern coal mines have only a dozen miners.

As if the coal mine itself weren't enough, the bus carrying our field trip breezed right past the monument marking the site of the Ludlow Massacre. In 1914, the Colorado National Guard and hired security guards burned and gunned a tent camp of striking mine workers. Twenty people, eleven of them children, were killed. The mine owners claimed that the fire was started by an overturned stove; Woody Guthrie's song said otherwise. In a small attempt to do better, each year for fifteen years I stopped my Princeton field geology trip at Manzanar—the only Japanese-American concentration camp to have an inmate riot—and explained the 1942 Japanese internment as a low point in American history.

Open-pit mines have much better safety records. They also cause major disruption of the surface. Open-pit mines are economic even when removing a foot of rock for every inch of coal. There are two attitudes toward reclamation:

- The government: The surface will be regraded to conform to the original land surface and planting will reestablish vegetation comparable to the original plant cover.
- Walt Disney: We'll put a lake over here, the golf course over there, pricey condos in between . . .

Red Cross members searching the ruins of the striking miners' tent camp after the Ludlow massacre in 1914. (Copyright © 1995–2004 Denver Public Library, Western History Collection, Colorado Historical Society, and Denver Art Museum)

Neither is necessarily correct, but as our need for energy becomes more desperate, we might get a bit more creative about "reconfiguring" the land instead of pure "reclamation."

Coal travels from mine to market in several ways. Water transport, by barges on rivers and by ships at sea, is the least expensive. Slightly more expensive is railroad transport. Transport by unit trains, consisting entirely of coal cars with automated loading and unloading, is only slightly more expensive than movement by water. Either water or rail transport can be accomplished for one or two cents per ton-mile. Ground-up coal can be moved as a water slurry in specially equipped pipelines, although the initial capital cost is high. And then there is coal-by-wire. If the end use of the coal is for generating electricity, why not put the electric generating plant at the mine mouth and ship the electricity on overhead power lines? Some large installations in the western United States are coal-by-wire facilities. This uncovers a deep philosophical division: Some coal-by-wire facilities add atmospheric pollution to previously pristine environments. One attitude says that we have very few unspoiled places left and they are to be protected; the alternative says that having a moderate level of air pollution everywhere is better than piling all the air pollution sources into the Los Angeles basin. I don't have an answer to this problem, but the debate is going to get more intense as we slide down the back side of Hubbert's peak.

In brief, coal is cheap, coal is versatile, and the major industrial economies have extensive coal deposits. In the face of the impending oil shortage, a possible game plan is 1) move electric generating capacity from natural gas to coal, 2) move the natural gas to power automobiles and trucks, and 3) save the remaining oil for aviation. However, the environmental problems with increased coal burning are not easily solved. Research efforts to solve coal's sulfur, carbon dioxide, and other environmental problems go back more than twenty years. We cannot pretend that coal's problems will be solved in a year or two by a crash research program.

Coal utilization is not a problem that we can ignore. Fantasizing about a fleet of nonpolluting fuel cell automobiles twenty years from now will not compensate for declining oil production in this decade. I hate to say it, but we likely will be forced to choose either increased pollution from coal or doing without a significant portion of our present-day energy supply.

Six

Tar Sands, Heavy Oil

Pitch, bitumen, tar, and asphalt go back to the dawn of civilization (all four names mean roughly the same thing). Moses's little basket and Noah's big ark were waterproofed with tar. Egyptian mummies were preserved with pitch. In fact, the word "mummy" comes from an ancient Persian word for bitumen. Most of these products came from oil, by both natural and artificial processing. However, there are nonpetroleum products that go by the same names: pine pitch and coal tar.

Natural sources include ponds and lakes of pitch: the La Brea Tar Pits in Los Angeles, the pitch lake of Trinidad, the asphalt pond above the Burgan oilfield in Kuwait. The tar can also come mixed with sediment, usually sand or sandstone. I visited one sand-asphalt quarry in Utah that was used occasionally by the local highway department. The standard two-lane blacktop Western highway consists of a mixture of sand and asphalt, usually the leftovers from oil refineries. In the Utah quarry, the ingredients were premixed and free for the digging.

Natural products cover the entire range from normal liquid oilfield oil, to thick oil that will still flow, to oil too thick to flow, to solid hydrocarbons. Production of thick "heavy" oil began before 1900 in

California. The Midway-Sunset and Kern River oilfields require artificial heating to recover the oil. Several methods were tried. A "fire flood" was inadvertently instigated by injecting air into some of the wells. Fire started spontaneously within the oil reservoir in the same way that oily rags catch fire. Unfortunately, sulfur in the oil was converted by heat and water into sulfuric acid, which in turn ate the steel pipe out of the recovery wells. Two less drastic methods came to dominate California heavy-oil production:

- Intermittent steam injection, known as "huff and puff." Typically, steam is injected into a well for two weeks. The well then produces oil—and water from the condensed steam—for a couple of months. When the production rate declines to a noneconomic level, another steam injection is performed.
- For steady injection, about half of the wells are used as steam-injection wells and the remainder are used to recover oil heated and pushed by the steam injection. The typical well pattern is a five-spot arrangement, named for the five spots on dice. The upscale word "quincunx" has the same meaning, but quincunx sounds too snooty for roustabouts.

The density of crude oil is an important indicator of its value; the lighter oil is closer to gasoline and commands a higher price. The ordinary measure of density is the mass per unit volume or specific gravity. However, crude oil density has its own scale, standardized by the American Petroleum Institute, known as "API gravity" or "degrees API."[1] (Because the API sets standards for almost everything in the oil business, my father used "that's about API" in a sarcastic sense, meaning that things were fouled up as usual.) Average crude oil tends to be around 25° API. Light, and more valuable, crude oils run in the high thirties, with an upper limit around 40° for gas condensates. The two largest oilfields in California, Midway-Sunset and Kern River, produce heavy oil with 13° API gravity. On the very heavy end, the density of water is 10° API. Oils below 10° API are heavier than water. When seeking oil this dense, the usual exploration tactic of looking for subsurface oil floating above subsurface water will not work.

Heavy oil, oil less than 10° API, also tends to have a high viscosity.

Viscosity reflects the resistance to flow of a liquid through a porous rock. High viscosity liquids have a high resistance and flow slowly; heavy oils can be ten thousand times more viscous than water. Again, the definition is backward to our intuition; I've gotten it turned around more than once. Below 10° API, the viscosity is usually high enough to make ordinary flow into a well too slow to be economic.

Tar-Sand Deposits

Although tar-sand deposits have been discovered in thirty countries around the world, two stand out by containing more oil than all the world's conventional oil wells. In the eastern part of Alberta, Canada, and just north of the Orinoco River in Venezuela are tar-sand deposits of mind-boggling size. The Canadians prefer to call their deposits heavy-oil sands. As a student, my first temptation was to explain them as former supergiant oilfields that came close to the surface where all the small (and economically valuable) molecules evaporated. The more likely explanation is that 90 percent of the oil liberated from oil source rocks never gets trapped in oil reservoirs; it simply migrates up to the surface. The Canadian and Venezuelan tar sands are near-surface sites where the unreservoired 90 percent from deeper in the basin approached the surface.

In the near-surface environment, several things can happen to crude oil. Natural gas and some of the lighter oil molecules can simply evaporate from the oil. Shallow groundwater can dissolve a few components from the oil. The biggest agents for change are bacteria, possibly assisted by some fungi. To the extent that oxygen is available, the favorite bacterial lunch is the straight-chain paraffin molecule. Even in the absence of free oxygen, other bacteria modify the oil. Eventually, the inedible leftovers are compounds containing multiple carbon rings. A continuous sheet of carbon rings makes up the mineral graphite. I get loud groans when I call these multiple-ring oil molecules "graffiti." Their proper name is asphalt, or tar, or bitumen, or pitch.

These ring compounds have a distinctive odor; they are called aromatics. If you walk by a crew heating tar to waterproof a roof, you get a strong whiff of aromatics. Several, but not all, of the ring compounds are known to cause cancer. The single-ring aromatic benzene is a legally recognized carcinogen in the United States. Despite that, benzene makes up about 30 percent of our commercial gasoline. Given current safety standards, if benzene were invented today, it would not be allowed in gasoline.

If you were so fortunate as to obtain a lease on an area of Canadian heavy-oil sands and figure out how to mine the sand and how to recover the heavy oil, you would still not have a marketable product. The oil viscosity is too high to flow through a pipeline. If you could mix the heavy oil with an equal amount of light crude oil, the pipeline (and the refinery) could handle the mixture. Unfortunately, the light crude oil comes from the far other side of Alberta. To make your lease profitable, you would need to set up part of an oil refinery at your oil-sand mine to convert the tar into smaller molecules that will flow readily, even in the cold of a Canadian winter.

The existing Canadian oil-sands mines carry out this initial refining step. The existing method breaks carbon-carbon bonds in the large oil molecules, but for chemical stability a hydrogen atom has to be attached at each of the newly exposed carbon atoms. In the past, the cheapest source of hydrogen was natural gas, but natural gas is now starting to be in short supply.

Athabasca, Alberta, Canada

First Nations (formerly Native Americans, formerly Indians) knew about the tar sands along the Athabasca River. The first scientific examination of the tar sands was published in 1820. Some of the deposits are right at the surface. If the depth to the oil sand is less than two hundred feet, the sand can be uncovered and mined economically. Oil from deeper sands has to be extracted through boreholes, called in situ extraction. Let's discuss borehole methods first, because they are close relatives to conventional oil wells.

Heavy oil can be moved by using either solvents or heat. Gas condensates, like propane and butane, have been tried as solvents but they are expensive. Heating, using steam, is the usual tactic. There are two steam methods for recovering extremely heavy oils. In Canada, the California huff-and-puff is renamed CSS, for Cyclic Steam Stimulation. The other technique is SAGD, for Steam Assisted Gravity Drainage.

CSS uses vertical wells drilled down to, and through, the tar-sand layer. A typical cycle is steam injection for a week, a week for the steam to warm up the oil, followed by a couple of months of oil production. There are two factors that can make CSS favored over SAGD:

- Mudstone layers, if present, favor CSS. Thin layers of mudstone (shale) deposited in between the sand beds can make horizontal flow much faster than vertical flow. As we shall see, SAGD is absolutely dependent on vertical flow.
- CSS is more flexible. Steam injection times and production cycle times can be adjusted by watching the well performance.

CSS currently dominates in Canada. Imperial Oil (an affiliate of ExxonMobil) produces 150,000 barrels per day.

SAGD was made possible by the evolution of slant-hole drilling. The ultimate in slant-hole drilling is turning the bottom of the well into a horizontal borehole that follows the productive sand. Imperial Oil developed and patented a procedure for drilling two horizontal holes, one about fifteen feet above the other. Steam is injected into the upper hole. Heated oil, plus condensed water from the steam, is recovered from the lower borehole. Several successful tests have been made with SAGD, but large-scale production is yet to come.

In situ production, including CSS and SAGD, has a lower environmental impact than mining but recoveries are quite low. Usually, less than 20 percent of the oil is recovered. However, because the total oil-sand resource is huge, 20 percent is more than welcome. Twenty percent of a big number is still a big number.

In Canada, about three-quarters of the heavy oil is recovered by surface mining. The mines get more publicity because of the awesome size of the equipment. The new Shell mine, which began producing in the first half of 2003, uses Caterpillar trucks that hold four hundred tons of oil sand. Bucyrus shovels scoop up one hundred tons in each bucket load.

In 1967, mining of tar sands was initiated by Sun Oil Company, the last of the large oil companies to be family owned. Although Sun

Truck in an Alberta oil-sand mine. It can carry four hundred tons of oil sand in a single load. (Shell Photographic Services, Shell International Ltd.)

sold out their interest in 1995, the mining operation retains the company name as Suncor Energy Ltd. Early on, oil sales from the Suncor mine exceeded the operating costs, so in that sense the mine was profitable. However, the capital investment initially was being returned very slowly. As late as 2001, operating costs "excluding capital recovery" were still being publicized by other companies.[2]

In 2002, Alberta heavy-oil sands were producing 33 percent of the Canadian oil production, a percentage that continues to rise. There is plenty of tar sand, equivalent to more than a trillion barrels of conventional oil, which sounds very promising and a reason, perhaps, to hold on to the family SUV. It's not that easy, however:

- Existing oil-sand operations are major users of natural gas. The energy necessary to produce either hot water or steam and the hydrogen needed for upgrading the oil most often come from natural gas. Before 1980, natural gas was a bargain-priced by-product of conventional oil production. As discussed in Chapter 4, this is no longer the case. Especially during the winter of 2002–03, there was a rise in natural gas demand and natural gas prices. New oil-sands operations, as well as existing facilities, will be much less profitable under high natural gas prices.

- Water, a major ingredient for both surface mining and in situ operations, isn't limitless. Most of the water can be recycled, but there is still a net removal of surface or subsurface water. In addition, environmental concerns are raised about the dispersal of water after it has been utilized in oil recovery.

- In surface mining, economies of scale are vitally important. Although one truck driver costs about the same whether driving a ten-ton truck or a four-hundred-ton truck, the capital costs differ dramatically. The gigantic scale requires a $2 billion to $5 billion investment before the first barrel of oil is delivered. This limits the number of possible participants. Further, it makes potential players cautious. Early in 2003, two companies delayed indefinitely their plans for new oil-sands mines.

- The infrastructure available for building new facilities is limited. It isn't polite to include welders and construction engineers as "infrastructure," but the total number of trained people to build major new plants is not easily expanded. Fort McMurray, a town of about thirty thousand located two hundred miles northeast of Edmonton, is the only sizable town in the oil-sands area. You can't construct a $4 billion facility by hanging out a Help Wanted sign in Fort McMurray.

My guess is that the CSS version of in situ recovery is the process most readily expanded on a short time scale. The incremental investment in a single well is small enough to allow a wide range of participants. Equipment and methods are not exotic. The biggest improvement would be redesigning the process to minimize dependence on natural gas.

Orinoco, Venezuela

Along the northern bank of the Orinoco River is the *"cinturón de la brea,"* the belt of tar. (For the culturally conscious, note that "La Brea Tar Pits" is redundant. It's like saying "Sierra Nevada Mountains.") Estimates are around 1.2 trillion barrels: comparable to Canada and more than all the world's cumulative oil production to date. Although interest in the heavy-oil belt dates back to the 1930s, actual production did not begin until 1988. By 1999, heavy-oil production in Venezuela was 54,000 barrels/day, about 10 percent of the size of Canadian production. Projects now under way are expected to raise the total to 500,000 barrels per day, but political unrest in Venezuela may result in delays.

All four of the heavy-oil projects in Venezuela are joint projects between major international oil companies and PDVSA (Petróleos de Venezuela). In both Venezuela and Canada, the oil sands are close to the surface. However, the warmer average surface temperature in Venezuela, as compared to Canada, makes drilled wells more appropriate than surface mines.

When initially recovered, Venezuelan heavy oil contains around three hundred parts per million of vanadium. Vanadium? Vanadium comes from hard-rock mines, and it is used mostly to make tough steel alloys. What's vanadium doing in my oil? Because crude oil comes from the thermal breakdown of plant and animal matter, there are "molecular fossils" in the oil, including recognizable leftovers from chlorophyll in plants and hemoglobin in animals. These molecules have similar flat-ring structures especially adapted to hold a metal in the center of the ring: iron in hemoglobin and magnesium in chlorophyll. In oil, these "fossil" chemical rings are still winners at holding metals. Copper, nickel, and vanadium turn up in these rings, in small amounts, in most crude oil. Heavy oils are the leftovers after the bacteria finish eating, and bacteria aren't fond of the metal-containing rings. Venezuelan heavy oils are particularly rich in vanadium.[3]

A serious problem arises if vanadium-rich oil is burned in a gas turbine engine. Gucky deposits accumulate on the turbine blades. Worse, vanadium can act as a catalyst and the vanadium-rich guck starts converting the sulfur that was in the oil to sulfuric acid, which—as we have discussed—corrodes metal parts. Instead of undertaking the diffi-cult and expensive step of removing vanadium from the oil, fuel addi-tives are used to dissolve the vanadium-rich guck and send it out with the exhaust.

The research arm of PVDSA, in cooperation with BP, came up with a clever way of producing and marketing Venezuelan heavy oil. During the 1980s they developed and began to sell an emulsion com-posed of 70 percent heavy oil, 30 percent water, and a small amount of detergent. The mixture, known as Orimulsion, can be shipped by pipeline and by tanker ships. It is burned in the boilers of conventional steam power plants; the water does not put out the fire. The oil-water emulsion could become the dominant heavy-oil export of Venezuela.

Upgrading Venezuelan heavy oil to a low-sulfur light oil was intro-duced by a joint venture between PVSDA and TOTAL. (TOTAL is a French company formed by the merger of Total, Fina, and Elf Aquitaine. Insiders pronounce it tow-TAAL; the same people pro-nounce Target department stores tar-ZHAY.) The Venezuelan heavy-oil upgrading is parallel to a similar process in Canada.

The Future of Heavy Oil

Heavy oil from various sources now amounts to roughly 8 percent of the world total. The resource is large and the upgraded oil is compati-ble with our existing automobiles, trucks, and aircraft. Intense pressure will arise for increased heavy oil production.

- Access of potential Canadian producers to water, natural gas, and construction capabilities should be monitored with the hope that a single shortage does not inhibit the expansion of current oil-sand production. Availability of natural gas at a guaranteed price may be the most difficult to arrange and the most crucial. Who wants to plow $4 billion into a plant and then have increased natural gas prices make the whole project uneconomic?
- There is lots of room for creativity in developing better in situ recovery. One ambitious project involves upgrading the oil in

the underground reservoir before it is extracted. SAGD has yet to fill its expected potential. One Japanese group suggests intermittent steam injection into the lower borehole, which is normally the recovery channel.

- Heavy-oil extraction is a major consumer of energy. To avoid consuming part of the valuable oil to power its own recovery, there is discussion about installing a 500-kilowatt solar energy facility at the Midway-Sunset field in California. The Canadian oil sands could use heat and power from a nuclear reactor.

- There is room for some hybridization. Generating hydrogen to upgrade the oil produces carbon dioxide, which is currently being sent up the smokestack. The carbon dioxide can be used, and recycled repeatedly, as an aid to recovery in conventional oilfields. The challenge is to back off and search for an optimum arrangement for the whole system.

It would be nice if heavy-oil production would increase seamlessly to make up for the Hubbert peak decline in conventional oil. In the best of all possible worlds, an increasing stream of heavy oil would avoid rearranging our energy supply chain. Heavy-oil production from California, Canada, and Venezuela is included in the production graphs in Chapter 3. The tar sands and oil sands are not being brought into production as fast as conventional oil, outside the Middle East, is declining. As the Middle East swings into its decline phase, a rapid and enormous investment in tar-sand facilities would be required. In my opinion, the preliminary steps to acquire government permits, investment capital, and construction capability are not going forward on a scale large enough to postpone the Hubbert peak.

Seven

Oil Shale

By now it's a tradition: All discussions of oil shale begin with, "It's neither oil nor shale." Oil shale is simply an immature source rock for petroleum, but one that has never been hot enough to enter the oil window. If natural processes have not cooked an organic-rich rock, we can still cook it (more about recipes later). There is a reason to try: In round numbers, one ton of good oil shale will generate one barrel of oil. At $25 per barrel of oil, that's $25 per ton of rock. Reasonably profitable gold ore contains about $12 worth of gold per ton. If oil shale is twice as valuable as gold ore, why isn't somebody getting rich? Unfortunately, some very large companies have lost money trying.

Origin

Oil shale is deposited in an environment that lacks oxygen to oxidize the organic matter. It isn't just a lack of oxygen; some clever bacteria can oxidize organic matter using sulfate from the water. The bottom water above the sediment has to be low in both oxygen and sulfate. All the world's petroleum source rocks, as well as oil shales, were deposited in uncommon environments devoid of life.

Curiously, these oxygen-free "dead" environments are famous in the study of ancient life.[1] Any dead body that falls into oxygen-depleted (and sulfate-depleted) bottom water gets preserved intact. Bones are not scattered by predators lunching on the meat. Soft tissues, which never show up in normal fossils, are sometimes preserved. These few places where the meat and skin as well as the bones are preserved are famous: the Burgess shale, the Solenhofen limestone, the Green River oil shale.

In normal sediments, where oxygen is available at the sediment-water interface, a variety of worms and creepy things dig through the sediment in search of something to eat. Before the worms start digging, there is bedding, obvious layers in the sediment. All sediments start out with bedding. It is almost impossible to pour sand into a tank full of water without generating visible layering. If there is no oxygen, and no life, the layering is beautifully preserved. No worms burrow through the layers. There are quite a few sedimentary rocks with layering and visible burrows. However, most sandstones and limestones have a "massive" look: no visible layering, no visible burrows. Typically, the massive sandstones and limestones have been so repeatedly burrowed that their texture is analogous to mashed potatoes. The strongest evidence for intense

Fossil snake, preserved in fine detail because scavengers could not live in the oxygen-free bottom water of the Green River lakes.

burrowing is no visible burrows and no visible bedding. The usual situation consists of normal, burrowed sediments thousands of feet thick enclosing a few layers ten to thirty feet thick that contain enough organic matter to be oil shales or oil source rocks.

Marine oil shales and oil source rocks are usually deposited in nutrient traps: an arm of the sea with enough freshwater input to cause surface water (instead of deep water) to exit the basin. Dead bodies, most of them microscopic, fall out of the surface water carrying their internal chemical nutrients with them. Nutrients build up in the deep water. If it is surface seawater leaving the basin and deep water entering, the basin is a nutrient trap. It is not surprising that sediments deposited in a nutrient trap contain not only organic matter but also nutrients. The presence of inorganic nutrients like phosphorous and nitrogen is an important clue to the origin of oil source rocks and their unheated equivalent, oil shales.

An ordinary garden-variety fertilizer bag will tell you the percent-

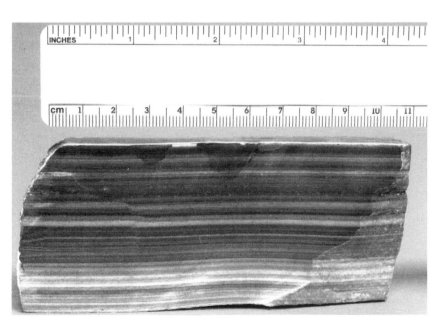

Thin layers within the Green River oil shale are probably annual layers, somewhat like tree rings. Preservation of the thin layering shows that bottom-dwelling burrowing animals were not present.

ages of nitrogen, phosphorous, and potassium that it contains. In seawater, potassium is abundant and available; nitrogen and phosphorous
are often in short supply. An example of a layer containing both organic matter and nutrients occurs in Wyoming, Montana, and Idaho.
The unit is named the Phosphoria Formation and is of Permian age
(260 million years old). Where the Phosphoria Formation has been
buried deep enough to fall within the oil window (7,500 feet to 15,000
feet), the organic matter has generated oil. About half of the oil produced in Wyoming came out of the Phosphoria. Around the town of
Dillon in southwestern Montana the Phosphoria never reached the oil
window, and as a consequence the most organic-rich layers are oil
shale. As for nutrients, there are major mines in southeastern Idaho extracting phosphate for fertilizer. Nitrogen is there too, in the form of
minerals containing ammonium ions, NH_4^+. In 1964, small amounts of
an unusual variation of the common mineral feldspar were discovered
in a hot-spring deposit in California. In place of the usual sodium, calcium, or potassium, the new feldspar contained ammonium. The new
mineral was named buddingtonite in honor of A. F. Buddington. Those
of us who studied under Dr. Budd thought he well deserved the honor,
even if the mineral was rare. Buddingtonite did not stay rare for long;
the Phosphoria sea was trapping nitrogen as well as phosphorous.
There are three billion tons of buddingtonite in the Phosphoria Formation.[2] In summary, the abundant phosphorous and nitrogen in the
Phosphoria Formation reinforces the hypothesis that the organic matter was generated and preserved in a nutrient trap.

History

And now we will present the award for the country that historically has
produced the largest amount of oil from oil shale. The envelope, please.
Tear, puff, and the winner is . . . Estonia! Why? Estonia had oil shale,
no oil production, and it used what it had. Other areas, like Sweden
and Scotland, had produced oil shale earlier. About thirty different
countries and eight states in the United States produced oil from
shales. When the 1901 Spindletop well near Beaumont, Texas, flooded
the world market with cheap oil, almost everybody, except Estonia,

gave up on oil shale. However, the view from Hubbert's peak is not generating an oil rush to Estonia.

The next award is for the country with the largest known deposit of oil shale. Tear, puff, and the winner is . . . the United States, for the Green River oil shale. Green River oil shale has been hanging over the conventional oil industry since I was a little kid. When oil was $3 per barrel, many people thought that if oil ever reached $8 per barrel, Green River oil shale would have its revenge on Spindletop and shut down the oil industry. Now with much higher oil prices, we had better take a closer look at the Green River.

More Than You Wanted to Know About Green River

The Green River oil shale deposits are big, very big. If the shale were heated, the amount of oil given off would be more than all the oilfields in the Middle East. The Green River deposits contain 60 percent of all the world's oil shale.

The Wyoming story says the oil shales were discovered when construction crews building the Union Pacific railroad back in 1869 made a circle of rocks around their cook fire and the rocks caught on fire. The Colorado story is more interesting. A fellow built a log cabin, including a fireplace and chimney made of an almost black stone he obtained locally. At his housewarming party he started a fire in the fireplace and burned down the fireplace, the chimney, and the log cabin. The "housewarming" wisecracks didn't help.

We're in the market for a Utah story, because the Green River oil shales occupy the area where Utah, Wyoming, and Colorado come together. The name comes, no surprise, from the Green River, which drains the area and eventually joins the Colorado River. It is not absolutely certain whether the Green River Formation was deposited in a single contiguous basin or whether it was formed in four separate basins.

As mentioned earlier, the Green River fossils are spectacularly well preserved: a snake, a turtle, even a bat. On one geology field trip I was taken to Hells Hole, Utah, where the specialty was fossil insects. The field trip stopped for an hour to let us all collect bugs. Particularly in

Wyoming, fish fossils are so abundant that thousands of them have been sold to tourists for a few dollars each. Fossil Butte is now a national monument.

The fossils tell us two additional things:

- Deposition took place during the last two-thirds of the Eocene epoch, about forty million years ago.
- The Green River deposit is nonmarine; it originated as sediment deposited in a lake (or lakes). There are no sharks, sea urchins, or other marine animals. The nonmarine part is a surprise; most oil shales and oil source rocks are marine.

Although the Green River lakes were not connected to the ocean, they were not freshwater lakes like the present-day Great Lakes. Think of today's Great Salt Lake, but ten times larger. And even then, the ancient Green River lakes and today's Great Salt Lake had radically different chemistries.

One of the classic papers that I assigned students to read was written by the geochemists Lawrence Hardie and Hans Eugster.[3] They explained the chemistry of lake waters in terms of "who runs out." For instance, as lake water or ocean water evaporates, the water will start to deposit calcium sulfate as the mineral gypsum, $CaSO_4 2H_2O$. Gypsum will precipitate until either all the calcium or all the sulfate is used up. It is highly unlikely that the original water contains exactly as much calcium (Ca) as it does sulfate (SO_4). Modern seawater, for example, contains more sulfate than calcium, so the gypsum precipitation ends when the calcium is essentially all used up and the remaining water is sulfate-rich. However, starting with more calcium than sulfate would branch the other way; a calcium-rich and sulfate-free water would result after evaporation. Hardie and Eugster drew a tree with limbs that divided each time that a component ran out. At the tips of the limbs, they placed the names of modern-day lakes with the appropriate chemistry. I find it interesting that the Great Salt Lake shares its limb tip with the world ocean.

The water in the ancient Green River lakes, back in the Eocene, was chemically unlike modern seawater. To a limited extent, the modern lakes in the eastern rift valley in Kenya and Tanzania are similar to the Green River lakes. Both are sources of commercially valuable amounts of sodium carbonate. About half of all the sodium carbonate mined in the world comes from the Wyoming portion of the Green River lakes. It goes into glassmaking, laundry products, paper manufacture, and a variety of chemical products. The stuff in the familiar yellow Arm and Hammer baking soda box comes from Green River carbonate.

The previous paragraph said "to a limited extent" because some aspects of Green River chemistry have no modern equivalent. At least nine minerals occur in the Green River oil shales that have been found nowhere else on Earth.[4] Five other Green River minerals are found elsewhere, but usually they come not from lake sediments but from unusual high-temperature rocks. For instance, leucosphenite can be found in the Green River Formation and in an unusual igneous rock in southwest Greenland, and nowhere else. In particular, the unusual Green River minerals often contain boron or fluorine, presumably telling us that the lake water contained important concentrations of these two elements.

To summarize: In a limited time (during the later part of the Eocene), in a limited area (about ninety miles across), a chemically weird oil shale that could produce more oil than the Middle East was deposited. It shouldn't come as a surprise that this unusual product came from an unusual setting.

For some years now, Eldridge Moores, a professor of geology at the University of California at Davis, and Yildrim Dilek, a professor of geology at Miami University in Ohio, have been pointing out that a mountainous area as high as the Tibetan Plateau existed in eastern Nevada during the Eocene.[5] Their assertion has not received a great deal of attention, in part because Moores bubbles up with so many interesting ideas. For instance, during the late Precambrian, Eldridge has Antarctica parked up against the western United States. But Moores is not a crank; a few years ago he was president of the Geological Society of America. For the moment, however, we will bypass the other clever Moores ideas and focus on eastern Nevada and westernmost Utah.

If a Tibetan plateau existed in Nevada and western Utah during the Eocene, the Green River lakes would have existed in an extremely dry rain shadow. The rain shadow that most of us know about is caused by the Sierra Nevada. Air in the prevailing westerly wind is cooled as

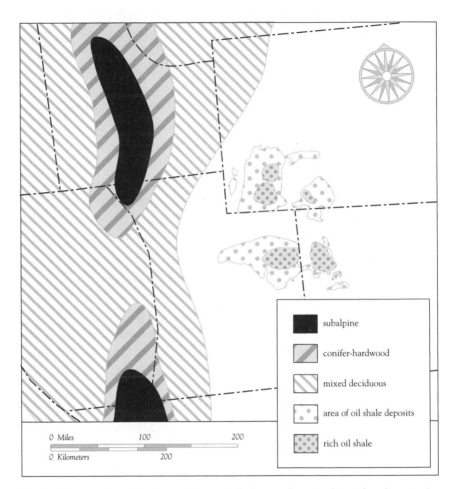

During Eocene time, the Green River oil shale was deposited in a dry climate. A major mountain range in Idaho and eastern Nevada is inferred from fossil plant leaves. At that paleolatitude, the prevailing wind direction would be from west to east, so a rain shadow would have existed in Colorado, Utah, and Wyoming, where the oil shales are now preserved. (Jeffrey L. Ward)

it is lifted up the western slope of the Sierra, depositing rain or snow. After the wind passes over the Sierra crest, nothing puts the water back, and much of the state of Nevada is sagebrush desert. During the Eocene, the area was at about 35° north, so winds from the west should have prevailed. If the Eocene plateau were higher than today's Sierra Nevada, the area east of the plateau would have been drier than a sagebrush desert.

Most of Nevada and the western half of Utah is named the "basin and range province" because it is broken into fault blocks. There is an unfortunate tendency to think of it as a single homogeneous province. However, the ranges and basins cut across some extremely diverse preexisting geology. The suggested high plateau in the Eocene occupied only the northeastern portion of the basin and range province.

Moores didn't just make up the idea of an Eocene Tibet while downing a few beers in a Nevada bar. The beginnings of his theory go back to his 1963 Ph.D. thesis on eastern Nevada; he mapped an area near the town of Ely. ("Near" in this case is sixty miles from Ely; there aren't any other towns in between.) The whole region around Ely contains giant faults having the geometry of low-angle gravity slides. Today, similar faults in the Himalaya—including one named the Main Central Thrust—are suspected of having the same geometry. It isn't subtle; the fault cuts across the upper part of Mt. Everest.

Over the years, Moores has been keeping a mental list of other evidence for an Eocene plateau in eastern Nevada.

- The paleobotanist Daniel Axelrod described Eocene fossil plants near Elko, Nevada, as having "alpine" affinities.
- The gold in the fabulously rich placer gravels of California may be from the Eocene-age plateau in Nevada. While the California gold rush was still in progress, G. K. Gilbert showed that the richest of the placer gravels was deposited by rivers *before* the Sierra Nevada was uplifted as a mountain range. Further, the gold-bearing veins in the Mother Lode were not the dominant source of the placer gold; the rich placers extended upstream from the Mother Lode veins.
- Extensive volcanic activity, otherwise unexplained, appeared across Wyoming, Montana, and Idaho during the late Eocene.

During the California gold rush, the gravels in the present-day rivers were depleted first. Then, older gravel deposits were excavated with giant hose nozzles called monitors. These hydraulic mines were closed in 1884 because the rivers were clogged with mud and silt, and they were starting to fill San Francisco Bay. (Culver Pictures)

The extensive volcanism associated with the Tibetan plateau is currently not well understood. A meaningful comparison awaits further study.

None of these by itself proves the existence of an Eocene high plateau in the eastern part of the basin and range province. Nevertheless, taken together, and in the absence of opposing evidence, Moores is more likely than not to be correct.

In 2002, I ran across a small peculiarity that may be part of the same puzzle. A former student and I tried to locate the source for some zeolite crystals that I had collected in 1958. We think we found the place, but in addition to the zeolite we found two crystal forms of the common mineral calcite that were unlike any that we had seen. We were—and still are—puzzled, but we realized that the rocks we were sampling were formed at the time (Eocene) and at the place (northeastern basin and range) where Eldridge Moores had his highest plateau. The crystals may have originally formed as something unusual in a cold, arid lake on top of the plateau and later recrystallized to calcite.

Let's entertain the hypothesis that the Green River lakes were in an extreme arid climate, with an unusual water chemistry and bottom water without significant oxygen. Anything from single-celled algae on up to snakes and turtles was preserved. An enormous deposit of organic matter accumulated. Either natural heat, associated with deep burial, or artificial heat will generate oil. How does the oil cooked out of the Green River rocks compare with ordinary oil?

The source rocks for most of the world's oil were formed in oceanic environments: marine source rocks. However, there are a few examples of crude oil from nonmarine rocks. The nonmarine oils tend to have a very large percentage of straight-chain hydrocarbon molecules. Marine source rocks tend to generate oil containing branched-chain and ring hydrocarbons in addition to straight chains. The straight-chain hydrocarbons line up nicely into crystals. The waxy paraffin used to seal up homemade jelly is an example.

Around 1973, the first delegation of mainline Chinese scientists visited the United States, and Princeton was one of their stops. We took them into our small geology museum to show them spectacular fossils from the Green River Formation. The lead geologist in the Chinese delegation said that China was producing some oil that came from nonmarine source rocks. Hollis Hedberg, a famous American petroleum geologist, started bouncing up and down, saying, "I knew it. I knew it." The Chinese had recently begun exporting oil to India, but news stories said that when the first tanker ship full of Chinese oil arrived in India, the oil had turned into a gooey solid. The tanker had to be unloaded by shoveling the oil out of the tanks. Hedberg knew that crystallizing paraffin at room temperature was the signature of a nonmarine source rock; he was delighted to have his guess confirmed.

A small part of the Green River Formation, mostly in Utah, has been buried into the oil window. The Roosevelt oilfield and several smaller nearby fields seem to be producing oil from Green River source rocks. These fields have produced five million barrels of oil and are still active. And, yes, the oilfields have severe problems with paraffin crystallizing out of the oil and plugging up the pipe.

If we multiply length × width × thickness × the amount of oil potentially generated per unit volume, we get 1,500 billion barrels. The remaining reserves in the Middle East oilfields are about 400 billion barrels. Of course, the exact number you get for the Green River oil shale depends on the lowest oil content that you think might be commercial. In one sense, right now none of it is commercial, so the answer comes out zero.

The highest oil content recoverable in the Green River is close to one barrel per ton. The richest unit is informally named the Mahogany Ledge, a reference to the rock's dark brown color. A lot of talented people have paid close attention to the Green River potential. The Colorado School of Mines ran annual meetings and published annual volumes from 1974 through 1986. The United States Geological Survey, the Bureau of Mines, and the Atomic Energy Commission all worked on the possibilities. Shell, Exxon, Union Oil, Sinclair, and Atlantic-Richfield set up pilot plants on the legendary Mahogany Ledge.

Could the problem be *that* hard? In one sense, it is easy. When I was teaching beginning geology I extracted oil from Colorado oil shale in a tabletop demonstration. I put a couple of tablespoons of oil shale, broken into quarter-inch chips, into a test tube and heated it. As oil started to be cooked out of the shale, I lit the open end of the test tube. The burning oil spread a nasty smell through the entire two-hundred-seat lecture hall. In addition to waking up sleepy students, the demonstration established that oil from oil shale was a high-sulfur, less-than-premium crude oil.

Oil Shale Recipes

In the traditional oil shale operations of the early 1900s, the oil shale was loaded into a large tank, called a retort, and the tank was heated to drive off the oil. One disadvantage: The leftover shale, after heating, fluffs up to a volume about 20 percent larger than the original shale. If you try to dispose of the spent shale back into the hole where you mined it, you have spent shale left over. Do you acquire a canyon and fill it with the extra shale?

Lots of variations on the simple retorting process have been examined. After the oil is cooked out, a carbon-rich char coats the mineral portion of the original shale. That unmarketable char can be burned to supply heat for the retort. Of course, I wonder whether the char could be converted to synthesis gas, along the lines of the Texaco coal process, and the hydrogen used to upgrade the oil.

Instead of mining the oil shale, toasting it in a retort, and throwing out the ashes, attempts have been made to break a volume of oil shale into chunks underground and heat it in place to recover the oil—in situ extraction. Some experiments have been tried, but they did not blossom into commercial production. The jury is still out on in situ.

In ordinary petroleum refining, in processing Canadian tar sands, and in producing oil shale, commercial viability involves the hydrogen-to-carbon ratio. The lower-grade crude oils and tar and shale oils have a lower hydrogen-carbon ratio than the gasoline that we crave. For upgrading these oils, there are two obvious hydrogen sources: methane (natural gas) and water.

Until recently, natural gas has been the cheapest source of hydrogen. Methane is currently used in upgrading Canadian tar-sand oil. However, natural gas prices have been going up rapidly. Also, as was explained in Chapter 4, our surplus natural gas resources are being used up for new electric-generating facilities.

Water is cheap, but it isn't exactly free. The Green River (this time meaning the river, not the oil shale) flows into the Colorado River. Las Vegas and San Diego have enormous thirsts for additional water, but Colorado River water is apportioned through a treaty with Mexico. Diverting a major amount of river water for oil shale processing is nearly guaranteed to generate litigation. The lawyers' torts might cost more than the retorts.

Cooking oil shale with steam has been tried.[6] The results are encouraging, but a commercial success has not emerged. One new item on the horizon: Separation of oxygen from air has become less expensive. One secret of the Texaco process for making hydrogen from coal and

A pilot plant for extracting oil from oil shale sits abandoned near Eagle, Colorado. This photograph was taken in 2003, when the plant had been closed for twenty years. The benches behind the plant are the result of mining oil shale. (David Zalubowski, Associated Press)

water is using pure oxygen instead of air. Air is three-quarters nitrogen. There are losses associated with warming and cooling the nitrogen, and the nitrogen oxides are unwelcome by-products of the process.

Is there anything left that could possibly be done to get American oil shale into production? Increased oil prices probably won't suffice. If I were going at it, here is what I would try:

- We badly need a by-product or coproduct, something made out of spent shale, or at least the extra shale that won't fit back in the ground. People have looked at glass, cement, and wallboard. Choosing something from that list, or finding another by-product, would be an enormous help.
- The sodium carbonate beds in Wyoming are interleaved with some pretty good quality oil shale. Is it possible to mine a mixture of not-quite-economic sodium carbonate and some not-quite-economic oil shale, find some clever by-product, and make a profit?

- Develop a computer model for a generalized retorting process. It isn't that the computer is a magic guru. Writing the program makes you think about each relevant issue. Once the program is even reasonably correct, you can start asking what-if questions much more cheaply than you could by building and abandoning retorts.

Here is an example of a successful computer program. Kennecott Copper was having trouble optimizing large heap-leach operations for recovering copper. Larry Cathles, then on the Kennecott research staff and now at Cornell, worked out a computer routine to model the flow of air, water, and heat, and the chemical actions of bacteria within the heap. The process became more effective, not so much because the computer program was of uncanny accuracy but because computer simulations told you what to avoid. My recommendation would be to start by hiring Larry Cathles and one Estonian.

Eight

Uranium

M. King Hubbert's famous 1956 paper predicting the peak in U.S. oil production had the title "Nuclear Energy and the Fossil Fuels." He saw nuclear power as the long-term replacement for oil and natural gas. Even if Hubbert turns out to be absolutely correct in his evaluation of oil production, that is no guarantee that his opinion about nuclear energy is correct. However, his message deserves a careful evaluation. Today, nuclear power is enormously unfashionable in the United States. The 1979 meltdown at Three Mile Island and the 1986 explosion at Chernobyl reinforced an underlying fear of nuclear radiation. No new nuclear power reactors were added in the United States after 1973.

Generating electricity from uranium does not add carbon dioxide to the atmosphere. To the extent that nuclear facilities are someday accepted, we have the uranium. We have an existing body of engineering experience.

From a public relations standpoint, it would be preferable to discuss nuclear power plants without ever raising the unpleasant subject of nuclear weapons. However, to get a clear overall perspective, we regrettably have to discuss bombs as well as generating electricity. The two histories, the two technologies, and the need to bring both under control are heavily intertwined.

All reactions, chemical or nuclear, can release useful energy only if the starting materials have a higher energy level than the final products. A graph of the nuclear energy level against the size of the nucleus is shaped like the bottom of a typical swimming pool: steep on one end and gradually shallowing toward the other end. The lightest nuclei, like hydrogen and helium, have high nuclear energy levels. The lowest level is for nickel and iron, and the energy rises slightly for very heavy elements up to uranium. Nuclear energy can be obtained either by splitting the largest nuclei into smaller pieces (fission) or by combining two small nuclei into a larger one (fusion).

Fission was developed first. In the years just before World War II, laboratory work showed that the uranium nucleus, after absorbing a neutron, would split into two new nuclei of roughly equal size. Although both pieces were larger than iron, they were lower than uranium on the binding-energy curve. Energy would be released, lots of it. In late 1938 and early 1939, experiments showed that neutrons would also be released and the process could become self-sustaining. Natural uranium contains two different nuclei: uranium 235 (less than 1 percent) and uranium 238 (the remaining 99 percent). In 1939, Niels Bohr and John Wheeler showed that it was the scarce uranium 235 that was splitting.[1]

I heard Wheeler lecture about the discovery. He made it sound easy. Nuclei with an even number of protons and neutrons are slightly lower on the binding-energy graph, odd numbers are higher. When U 238 takes in a neutron, it momentarily becomes U 239, which is odd, and we can think of it as soaking up energy. On the other hand, adding a neutron to U 235 makes it even. For a brief moment, the extra energy is expressed as vibrations of the uranium nucleus, which then splits into two fragments.

With World War II looming, Leo Szilard and Eugene Wigner got Einstein to write the famous 1939 letter that told President Roosevelt that nuclear weapons were possible and warned him that whichever nation developed them first would have a substantial military advantage. Two possible routes to a bomb emerged: separating U 235 out of the natural uranium mixture and generating a new chemical element, plu-

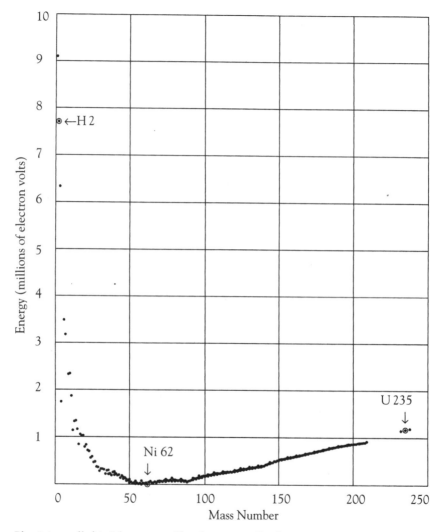

Physicists call this "the curve of binding energy"—the energy in an atomic nucleus divided by the number of protons plus neutrons. The most stable nucleus is nickel 62, shown circled at the bottom. Heavy elements, like uranium 235 (circled on the right), and light elements, such as deuterium (hydrogen 2, circled on the left), release huge amounts of energy if they are rearranged as nuclei closer to nickel 62. Traditionally, this diagram is drawn with the vertical scale reversed. I prefer the concept that energy is released when things flow downhill.

tonium, in a nuclear reactor. Both routes worked: A uranium bomb destroyed Hiroshima and a plutonium bomb wiped out Nagasaki. It should be remembered that at the time they were built, neither the Hiroshima nor the Nagasaki bomb was guaranteed to work. Either could have been reduced to a fizzle if a stray neutron from inside or from outside the bomb initiated the nuclear reaction halfway through the final explosive sequence. Under wartime pressure, both designs were developed.

The plutonium route required building nuclear reactors. The first test reactor, built beneath the football stadium at the University of Chicago, showed that the controlled release of nuclear energy was possible. After the first successful test run, Eugene Wigner produced a bottle of Chianti, which he had purchased in a wine shop in Princeton.[2] He was proud of his insight to stock up on European wine before the war started; by comparison, he claimed that designing a nuclear reactor was easier. Wigner, who had been trained as a chemical engineer before he took up physics, designed the large reactors for plutonium production at Hanford, Washington. Our present-day nuclear power reactors are of a very different design; the Chernobyl disaster involved a design similar to the Hanford reactors. It is a tribute to Wigner's genius that the Hanford reactors did not have a serious accident before the last one was retired in 1986.

Even as the nuclear weapons were being designed at Los Alamos, Edward Teller was looking at the other end of the binding-energy curve. The steep slope at the light end of the curve showed that even larger amounts of energy could be released. The possibility was known as a hydrogen bomb or "the Super." Up through 1950, both Teller in the United States and Andrei Sakharov in the Soviet Union pondered a number of unsuccessful designs. Although the details are still highly classified, apparently it was the mathematician Stanislaw Ulam who suggested to Teller that X-ray photons from an exploding fission bomb could be focused in order to heat and compress a mass of light elements.[3] The heat and density brought hydrogen nuclei close enough to fuse. Both the Americans and the Russians were soon conducting absolutely frightening atmospheric tests of multimegaton "hydrogen" bombs. Actually, the target materials were not garden-variety hydrogen. Hydrogen nuclei with extra neutrons, as well as the light element lithium, supplied the energy.

If nuclear fusion makes a hydrogen bomb possible, could we build a hydrogen-fueled fusion reactor? In 1951, an idea for a fusion reactor occurred to Lyman Spitzer while he was halfway up the Aspen ski lift. (His family later complained that it was the worst vacation they ever took. Spitzer had the idea on the morning of the very first day of their vacation; he got off at Midway, skied down, and spent the rest of the vacation writing in the hotel room.) Because Spitzer was an astronomer, he was well aware of the discovery by Hans Bethe that the sun is powered by nuclear fusion. Spitzer considered using magnetic fields to confine an electrically heated gas hotter than a million degrees. Spitzer's insight led to an experimental search to develop a controlled-fusion power reactor, an effort that has continued for fifty years. During the early years, experimental fusion reactors consumed a lot of power, and the only reward was a few neutrons coming out to

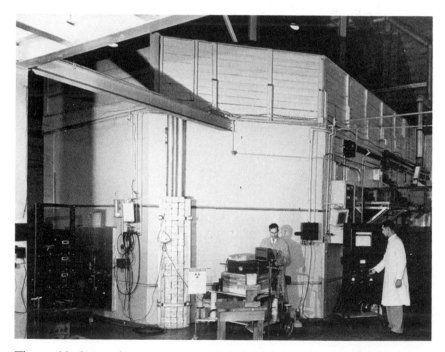

The world's first nuclear reactor was built in 1942 at the University of Chicago. The reactor consisted of unenriched uranium rods interspersed with blocks of high-purity carbon (graphite). (Bettman/CORBIS)

show that something was happening. As the experiments got bigger, their power supplies became larger than my three-story house. Although more energy was being produced, the energy output was smaller than the energy put in. An international project called ITER is now planning to get close to as much energy out as goes in, but it will take at least twenty more years before a commercial fusion power reactor is expected.

A nuclear-fission weapon, formerly called an atom bomb, has at its core a mass of plutonium or almost-pure uranium 235. In order to explode, the fissionable mass has to be large enough and dense enough so that most of the neutrons liberated by the initial fission will be captured by other plutonium or uranium nuclei, which then fission in turn. In all but the most primitive weapons, a shell of conventional chemical high explosive is used to compress the fissionable pit from a stable

In 1951, Lyman Spitzer Jr. started the program to generate energy by fusing hydrogen nuclei. In 1946, before there were artificial satellites, he wrote about the advantages of putting a telescope in space; he was the father of the Hubble Space Telescope. This photo was taken in 1989, when Spitzer was receiving yet another well-deserved award. (Princeton University Archives)

Experimental steps toward fusion reactors have gotten larger with time. The first experiments in the 1950s were large tabletop devices. This is a 1956 machine at the Princeton Plasma Physics Laboratory. The hot gas is confined in an oval horizontal tube, about six inches in diameter, at the center of the spool-like coils. Experimental machines today have confining tubes larger than six feet in diameter, and an experimental reactor occupies its own building. (Princeton University Archives)

size to a density where the neutron multiplication can happen. Part of the engineering involves the absence of stray neutrons until the pit reaches its maximum density, and then supplying a small number of neutrons to start the chain reaction.

As mentioned above, a fusion weapon, or hydrogen bomb, contains a minimum-size fission bomb, known as the primary. When the primary explodes, a messy mix of burned chemical explosives, fission products, neutrons, X-rays, and vaporized hardware comes rushing out. Out of all that mess, only the X-rays move at the speed of light; everything else is slower. The Ulam-Teller invention lets the X-rays race ahead of the rest of the flaming garbage. Those X-rays are focused to compress a secondary target containing atoms whose nuclei can fuse and liberate energy.

There is an important conclusion from this long and terrifying story about bombs and reactors. Almost all modern nuclear weapons are hybrid fusion-fission bombs. A uranium or plutonium fission bomb has tritium (hydrogen with two neutrons in the nucleus) added to supply extra neutrons to make sure that virtually all of the uranium (or plutonium) participates. In the secondary, or fusion, part of a hydrogen bomb, a slug of uranium is included. The bomb would work without uranium in the secondary. I am told that the uranium simply generates more bang for the buck. There is a parallel in the world of power reactors. Fission reactors are typically starved for neutrons; fusion reactors have neutrons coming out their ears. I have been grousing for twenty years that the only way a fusion reactor is going to make sense is as a component of a fusion-fission hybrid. Fusion scientists, however, don't want to hear about hybrids. Their beautiful pure-fusion reactors are "clean" in the sense that they do not generate long-lived radioactive fission products. They don't want their nice reactors messed up with dirty uranium.

Fission Reactors

The fission part of a nuclear reactor simply generates heat. If you want to propel a nuclear submarine or generate electricity, you use that heat to do work. Any system that converts heat to useful work is called a heat engine. In 1824, Sadi Carnot published a surprising result: The maximum possible efficiency of a heat engine depends only on the temperature of the heat source (hot) and the temperature at which the heat leaves the system (cold).

$$\text{Efficiency} = \frac{T_{hot} - T_{cold}}{T_{hot}}$$

where T_{hot} and T_{cold} are absolute temperatures, measured up from absolute zero.[4]

Carnot's result does not depend on the material inside the engine or its plumbing. Of course, all real engines have friction and other inefficiencies, but none of them can be more efficient than Carnot's ideal. (I was pleasantly surprised to find streets named Carnot in several French towns. It turns out that there was a famous family named Carnot—one was the president of France—no easy way to tell which one was being honored.) Not all systems that do useful work are heat engines. Fuel cells, your muscles, and electric motors are not heat engines and are not limited by Carnot's expression for efficiency.

Besides heat from a hot source, an electric-generating plant needs a cold place to dump the waste heat. Carnot's equation says the colder the better. In cold climates, an enlarged version of an automobile radiator can be used to release heat directly into the atmosphere. In warmer places, water can be sprayed onto the heat exchangers and the waste heat goes to evaporate the water. Of course, in many places the water is already committed to other uses and evaporative coolers cannot be used. Another solution is to locate the power plant alongside a river and run river water through to take up the waste heat. As much water comes out of the plant as goes in, so the power plant isn't using up any water. Right? Wrong! Part of the heated river water evaporates to cool the river back to its original temperature; the water loss is the same as direct evaporation inside the power plant. Locating the power plant at the seashore and cooling it with seawater may create some biological problems, but it does not use up the freshwater supply.

As nuclear power plants were being designed in the 1950s, some nuclear engineers predicted that electricity would become "too cheap to meter." Uranium was abundant, and it was inexpensive in the sense that the uranium contributed very little to the price of nuclear electricity. The major expense was the capital cost of building the reactor. Because of the capital costs, cheap electricity never appeared. More than 50 percent of the electricity I pay for in New Jersey comes from nuclear plants, and I pay about 11¢ per kilowatt-hour. What is interesting is that telecommunications have become too cheap to meter. In the 1950s, overseas telegrams cost 20¢ per word. Today I can shoot long e-mail messages to Tokyo and my monthly bill is a flat rate. Dartmouth College has stopped billing students for long-distance telephone calls.

There are many possible designs of fission reactors. About twenty basic types have been considered, and at least one experimental reactor of each type has been built. There is no need to list them here, although we should note that the variations involve three factors: fuel types, something to slow neutrons down (the moderator), and the means for taking useful heat out of the reactor (the coolant). The simplest fuel is natural uranium. The first experimental reactor under the University of Chicago football stadium, the World War II plutonium production reactors at Hanford, and the disastrous reactor at Chernobyl all used natural uranium interspersed with blocks of carbon as the moderator. The only successful modern design fueled by natural uranium is a Canadian design that uses heavy water (deuterium oxide) as a moderator. The Canadian-deuterium design, known as CANDU, has been used in seventeen electric-power reactors in Canada, and ten have been sold overseas. In a surprising move, Atomic Energy of Canada Ltd. is now promoting a new model (ACR-700) that abandons the CANDU design.

The use of heavy water (deuterium oxide) in the Canadian CANDU nuclear fission reactor allowed the use of unenriched natural uranium as fuel. A second advantage was being able to replace individual fuel rods while the reactor was running. The reactors are described as "a plumber's nightmare" because heavy water costs $400 per pound, and even tiny leaks cost money. (Candu-Ottmar Bierwagen/Spectrum Stock)

The next step up (or down, if you wish) involves reactors that would just sit there cold if fueled by natural uranium. The active ingredient in uranium is U 235, whose abundance is 0.7 percent. Enriching the U 235 content to 3 percent makes it possible to use ordinary water as both the moderator and the coolant. Ordinarily, separating U 235 from U 238 is a daunting and expensive task. Under wartime pressure, those wonderful folks who brought you Hiroshima developed gas diffusion for the separation. Today, gas centrifuges are cheaper and consume less energy. Commercial reactors generating electric power in the United States are of two slightly different designs. About a third of the reactors use heat from the uranium core to boil water; the resulting steam goes through a turbine to generate electric power. Two-thirds of the reactors heat pressurized water, which is not allowed to boil. In a heat exchanger, the heated pressurized water turns unpressurized water into steam, which in turn generates electricity. Some of the design choices may result from a peculiar piece of history. The pressurized water reactor was developed first for nuclear submarines, under Admiral Hyman Rickover's leadership. Early American power reactors were adaptations of the submarine reactors.

Efficiency comes from having the reactor core as hot as safety will allow and having the produced steam condense back to water at the lowest available temperature. Carnot's equation still applies. Designs for very high temperature reactors, cooled by molten metal, are on the drawing board.

After the core of one of the nuclear power reactors at Three Mile Island melted down in 1979, building new nuclear power plants in the United States came to a halt. Power plants already on order were canceled. The last order for a nuclear power plant that was actually constructed and put on line was placed in 1973.

In addition to reactors for electric power generation, a variety of special-purpose reactors have been built. One particularly interesting design was a line of research reactors called TRIGA, built by General Dynamics. TRIGA reactors were not just idiotproof but also professorproof.[5] The reactor's safety was not just a feature of the TRIGA control electronics; it was a property of the internal physics. If the reactor produced energy too rapidly, thermal expansion of the water slowed the nuclear reaction. Dozens of the TRIGA reactors were installed around the world; none of them have been involved in nuclear accidents.

The purpose of the World War II–era reactors at Hanford was to capture neutrons from U 235 fission in the more-common U 238 nuclei to produce the artificial element plutonium. The bombs exploded during the 1945 test at Alamogordo and over Nagasaki were plutonium bombs. Chemical separation of the reactor-produced plutonium from the fission products was another daunting task. The fuel elements, as they came from the reactor, were intensely radioactive; less than a minute of human exposure would be deadly. All the chemical manipulation had to be by remote control, and if anything broke there was no way to go in and repair it. The clever chemical step entailed finding a molecule that would bond selectively to plutonium and adding a hydrocarbon tail to the molecule so that it would dissolve alternately in water and in oil. This chemical separation was carried out in chains of shielded tanks called mixer-settlers; each mixer had a stirring paddle as the only moving part.

Today, the means to separate plutonium is commercially available, but it's intensely controversial. Plutonium recovered from used uranium is a low-cost fuel for generating additional electricity. However, the plutonium itself is lethally toxic, and rumor has it that building an illicit nuclear bomb from plutonium is particularly easy. In 1999, a Japanese power company had some of its spent uranium reprocessed in England, and the resulting plutonium was returned by ship to Japan. It caused intense nervousness around the world, especially because the ship might be hijacked by terrorists. Because the world has enough uranium ore, we probably will not have to separate plutonium commercially for at least fifty years.

The extreme version of a plutonium-fueled reactor is called a breeder reactor. In a breeder, plutonium is not just a by-product. The reactor is optimized to produce plutonium efficiently, and heat (and therefore electricity) is the by-product. Optimization involves finding materials that do not absorb neutrons; the goal is to have as many neutrons as possible absorbed instead by U 238. Hubbert's 1956 paper referred to breeders as the "catalytic burning" of U 238. In ordinary chemistry, a catalyst is something that increases the speed of a reaction without itself being consumed. In a breeder reactor, plutonium is consumed, but new plutonium

Superphénix in France, at 1,200 megawatts, is the world's largest breeder reactor. During years of testing, Superphénix was shut down for repairs more than half of the time and was closed permanently in 1996. (Thierry Saliou, Associated Press)

is generated by neutron capture. It is the common U 238 that is progressively consumed; plutonium is the catalyst.

A small American breeder reactor at Fort St. Vrain in Colorado operated from 1973 to 1989. In a more ambitious program in France, larger breeder reactors named Phénix and Superphénix were built. Superphénix was shut down in 1996; estimates of the cleanup costs for it exceed a billion dollars.

There is a parallel between the plutonium breeder reactor and the "clean" fusion process being developed. Fusion utilizes naturally occurring heavy hydrogen, known as deuterium, as fuel. However, the other reactant is double-heavy hydrogen, called tritium. Tritium occurs naturally but in incredibly tiny amounts. Fusion reactors will have to generate their own tritium. Absorbing the fusion-produced neutrons in a blanket of lithium serves two purposes: replacing tritium and recovering the energy produced by fusion. In that sense, the fusion reactor is a breeder. Tritium is the catalyst, deuterium is the fuel.

The Uranium Supply

Although uranium had been known to chemists since 1789, there was no real market for the stuff. Small amounts have been used since Roman times to give a fluorescent yellow color to glass. The only commercial source of uranium was as a by-product of mining other metals. Before 1900, the largest source was a uranium-silver mine at Joachimsthal in Czechoslovakia, in the part that is now the Czech Republic. That mining district gave its name to silver coins, called thalers, which in turn became our word "dollar." When I was lecturing, I loved telling the students that the name for the almighty American dollar came out of a uranium mine in Czechoslovakia. It generated a whole roomful of skeptical looks.

In 1898, when Marie Curie separated radium and radon from Joachimsthal uranium ore, the by-product took on a new importance. Radium and radon were part of the series of natural radioactive elements that slowly transform uranium into lead. Both radium and radon were found to be useful in treating cancer. Two additional uranium mines came into production; the by-product was now the product. One was the Shinkolobwe mine in southern Congo, the other was the Great Bear Lake mine in northern Canada. Small amounts of uranium were also produced in the Colorado Plateau area of Utah as a by-product of mining vanadium.

The American effort to build nuclear bombs during World War II was hindered by the lack of a domestic uranium supply. However, a timely 1,250 tons of high-grade Shinkolobwe uranium ore turned up in steel drums in a Staten Island warehouse. M. Edgar Sengier, managing director of the mining company, imported the uranium ore with the expectation that it would become valuable.[6]

After 1945, and through much of the Cold War, there was a major push to find uranium ores in the United States. Before then, whenever a pickup truck slowed down to watch a bunch of geologists examining a road cut, the question was, "What you guys looking for, gold?" From 1945 to 1985, the question was, "What you guys looking for, yewrainium?" After 1985, the price of uranium went down, gold prices increased, and we really were looking for gold.

Exploration for uranium was easier because of its radioactivity. The radioactivity of pure uranium by itself consists of alpha particles that can be stopped by a single sheet of paper. On the other hand, natural uranium ores contain a whole series of radioactive elements: steps in the decay of uranium to lead. Several radioactive steps within the series emit gamma rays that can penetrate several feet of rock. A handheld radiation counter was an enormous help in the search; for a while it seemed there was a counter in every pickup truck in the western United States. Even better, counters in low-flying aircraft could scan enormous areas. For prospectors, the plane could become a mixed blessing. Early on, when a plane discovered a radioactive hot spot, the pilot would fly back and forth across the anomaly to see how big it was. The plane would then land at the airport and the crew would go out to establish their mining claims. When they arrived, however, they often found fresh claim stakes. People on the ground had seen the plane zigzagging around. Survey aircraft quickly learned to keep flying in a straight line even if they crossed a hot spot.

In addition to surface-radiation meters, similar radiation counters can be lowered on a cable in a borehole. In the petroleum industry, virtually every borehole is surveyed with a wireline radiation meter. The purpose is not the search for uranium; variations in the small amount of natural radioactivity are useful clues about the rock types. In the Paradox Basin in Utah in 1959, I discovered an enormously high radioactive reading on an oil-well log, probably from a uranium deposit. It was more than a mile below the surface, too deep to mine commercially. I kept my day job.

The geochemistry of uranium involves the interaction between two different levels of oxidation: U^{+4} and U^{+6}. The +6 is the more oxidized; the +4 is reduced. Under present-day atmospheric conditions, U^{+6} holds hands with oxygen and carbonate to form a water-soluble complex. Most of the uranium mined in the western United States came from interfaces between oxidized and reduced conditions. Here's how it works. Surface water, carrying dissolved oxygen and tiny amounts of soluble uranium, flows through a sandstone until it encounters bits of organic material in the sandstone. The organic material gets oxidized, using up the oxygen. As a response, the uranium changes from the soluble +6 complex to the insoluble +4 state. Over geologic time, commercially mineable amounts of uranium accumulate at the interface between oxidized and reduced uranium. Exploration for these deposits is systematic because the oxidized part of the sandstone is buff colored, while the reduced part is typically gray because of the organic material. If your drill hits gray sandstone in the eastern part of your exploration area and finds buff in the western part, you know the good stuff is in between. Choose a place roughly halfway in between and drill again. If you get gray, go west. Buff, go east. Keep subdividing it. Eventually, gotcha!

Where uranium was deposited by water flowing through sandstone, there is the possibility of extracting the uranium by flowing chemicals through the same sandstone (in situ extraction). No underground mine, no miners exposed to radon, no heavy equipment. One uranium mine in South Texas looked like a bitty toy oilfield: wells, pipe, tanks, each about a tenth the size of a regular oilfield.

The early earth's atmosphere did not contain free oxygen. As a result, the +4 oxidation state of uranium was the only stable form. There was no conversion to +6 and no water-soluble uranium. In that environment, grains of the mineral uraninite (UO_2) could travel in river sands as solid inert grains. Uraninite grains are heavy and would segregate in much the same way that heavy gold grains accumulate in placer deposits. This became of huge importance in South Africa. The Witwatersrand gold deposits, discovered in 1886, have produced more than half of all the gold ever mined in the world. The gold ores also carry uranium. In 1961, good evidence was published showing that the Witwatersrand deposits were gold placer deposits, but they had accumulated 2.7 billion years ago.[7] Earth's atmosphere did not then contain oxygen and the resulting ores were uranium placers as well as gold placers. During the 1960s, the price of uranium went high enough that some South African mines made more money from uranium than they did from gold.

One of my friends told the story of working on a drilling crew looking for an extension of the Witwatersrand ore deposits. As the gold values in their drill cores got higher and higher, the company management stopped informing the field crew about the gold analyses. However, the field crew had a radiation counter that they could lower on a wire into their boreholes to check on the uranium. Since they already knew roughly how much gold corresponded to a given amount of radiation, they could map out the subsurface deposit on their own. What to do with the information? Buy building lots in nearby towns; a major mine would soon appear.

Around 1975, a new and unexpected type of uranium deposit was discovered, in Australia and Canada. The deposits were in sandstones, like the ones found in the American West, but instead of being at an oxidation-reduction boundary, they were tucked in at the base of the sandstone. Also, instead of appearing in the midst of a stack of sedimentary rocks, the new deposits were specifically in the bottom of a sedimentary stack, right where the sandstone was in contact with much older underlying rocks. In geological lingo, the contact with much older rocks was an unconformity and the newly recognized uranium deposits are called unconformity-type deposits. Today, thirty years later, the origin of unconformity-type uranium deposits is still not well understood. But they are big and they have high-grade ores. Between them, the Canadians and the Australians produce 90 percent of the world's uranium.

The McArthur River mine in Canada contains ore that averages 21 percent uranium oxide.[8] There was no practical way to protect miners from that level of radioactivity. So the Canadians found a clever way of using an existing piece of equipment to conduct unmanned mining. The gadget of choice is a raise borer. The machine drills a ten-foot-diameter hole up through the ore and the chips fall onto a conveyor belt below. No miners even see the ore. It's not the cheapest way of mining, but with extremely high-grade ore it is still profitable.

The Case of the Breeder Reactor

In 1972, the Club of Rome's *The Limits to Growth* set many people to thinking about future constraints on industrial civilization.[9] However, there was little or no geological information in the book. In a narrower study, the U.S. Department of Energy released a 1978 report stating that breeder reactors would become necessary because of limitations on the natural uranium supply. Again, there was little or no geology in the DOE study. They were just plugging algebraic curves into a computer, and if the input curve was steeper than a certain level, the breeder reactor was needed. As you know by now, I'm in the business of selling geology. When I was a geology undergraduate, a frequently heard phrase was, "The present is the key to the past." Now, as a retired professor, I hear, "The present is the key to the future." All policy wonks and all futurologists need to keep a geologist around. Hubbert's peak is just one of the geological constraints on our future society.

There was even some doubt about what the uranium distribution would look like. How much uranium could be recovered, at each stage, as the uranium ore grades get leaner and leaner? Two possibilities have been suggested, one from Yale and one from Princeton. The choice between the two suggestions has major implications for the development of industrialized society. On a smaller, but still important, scale, the difference between the Yale and the Princeton models determines when or whether a breeder fission reactor is needed.

Brian Skinner, a distinguished economic geologist at Yale, posited that graphs showing ore grade versus tonnage for many metals would exhibit a double-topped curve.[10] The larger bump would give the distribution of the metal in ordinary rocks. Lead, for instance, has a size and electric charge similar to calcium. Ordinary rocks contain large amounts of calcium, usually several percent. Small amounts of lead can "hide" by occupying spaces normally reserved for calcium. However, Skinner pointed out that in economically mineable ores, the valuable metal is not hiding. The ore contains mineral grains with the metal as major constituent. For instance, lead ores almost always contain lead sulfide, the mineral galena, PbS. Instead of substituting in tiny amounts, lead is the major component in the mineral. Skinner then gave his opinion that minerals with tiny amounts of lead and minerals with major amounts of lead are two entirely different beasts and there is nothing in between. The consequences for society would be dramatic: Once the high-grade minerals were gone, the only way to get lead would be to mine common rocks, like granite, to extract the tiny amount of lead. Skinner gave his original article the intriguing title, "A Second Iron Age Ahead?" His inference was that our use of rare chemical elements like copper, zinc, and chromium would diminish and we would have to rely on geologically abundant metals like iron and aluminum. Skinner felt that his model was quite general and would apply to a wide range of metals, including uranium.

In contrast to Brian Skinner's view, my attitude is that everything looks like a bell-shaped Hubbert curve until proved otherwise. However, Brian explained to me an exceedingly important point: We can't solve this problem by making direct measurements. Suppose that I, as a young scientist, had developed a machine that could make one chemical analysis for zinc every second, night and day, 24/7. I go out the back door of the Princeton geology building, head roughly toward Seattle, and collect one rock sample per second. Skinner showed me that I would reach retirement age before analyzing zinc in enough rocks to tell whether he was right. Here is a way of understanding Skinner's

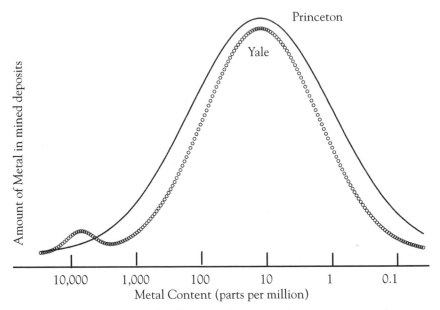

Brian Skinner, a professor of geology at Yale, proposed that uncommon elements in the earth's crust (such as copper) would exhibit a double-peaked distribution, as shown by the open circles in this graph. The high-grade, economically mineable deposits are the smaller peak on the left. I wondered whether the high-grade deposits were simply the tail of a single bell-shaped distribution. This diagram is just a cartoon to illustrate the concept. If it were drawn to scale, the ore deposits on the left would be too tiny to see. These bell-shaped curves are *not* Hubbert curves. Hubbert's horizontal axis is time. On this graph, and the following graph, the horizontal axis is a logarithmic (1, 10, 100, . . .) depiction of the metal concentration.

point: How many random rocks would I have to pick up before I discovered a commercial zinc mine by accident? Probably more than I
could analyze in one lifetime. Geologists who discover commercial zinc
deposits use informed hunches and especially they use analogies to
known zinc mines. There is no way to test the Yale-versus-Princeton
view by straightforward measurement.

Uranium offers an interesting test case. A low-flying plane with a
big radiation counter does make one rock analysis per second. (In addition to uranium, there are only two other chemical elements with
substantial amounts of natural radioactivity: thorium and potassium.
Some radiation counters are rigged to ignore thorium and potassium.)
A million feet of petroleum boreholes are scanned with wireline counters each year. Combining the gigantic library of radiation counter
records with the uranium content of mined ore deposits offers a unique
opportunity for testing between the two-peak Yale model and the
single-peak Princeton model. In addition to the intellectual interest,
the uranium distribution determines the need for the breeder reactor.

If history were nice and orderly, mining would begin by digging up
the richest ores and then turn progressively to the lower-ore grades. For
copper, this is roughly true. For uranium, it is not. The leanest uranium
ores ever mined in the United States were extracted during the years
immediately after 1945. In order to reconstruct the geologic curve of
the amount of uranium available at various ore grades, I needed to ignore the date the ore was mined and instead sort the uranium mining
history into different ore grades.

From 1945 through the 1960s, the only legal purchaser of uranium
in the United States was the Atomic Energy Commission, which was
later folded into the Department of Energy. Records of the ores mined
were supplied to the AEC with the agreement that individual company
records would remain confidential. In 1975, I applied to the DOE for a
research grant to examine the mine records in an effort to fit them into
the larger picture of uranium in Earth's crust. My application asked for
money and access to the records. The money turned out to be the
easy part.

By 1975, the AEC and DOE records had been placed on computer
tapes, inside a limited-access computer room, inside a one-story office
building, inside a tall barbed-wire fence on the edge of Grand Junction,

Colorado. In order to get inside the barbed-wire fence, one had to be fingerprinted and undergo an FBI background check. My two colleagues were Ian MacGregor, a geologist on leave from the University of California at Davis, and Jim Kukula, a Princeton undergraduate physics major and an expert computer programmer.

In the summer of 1976, Ian and Jim went to work in Grand Junction. The federal bureaucrats explained to them that the confidentiality agreement between the government and the mining companies specified that any published statistic had to include data from three or more companies. In a few instances, the DOE released data with only two companies represented in a statistical category. They got angry calls from companies that had subtracted their number from the total to figure out the other company's confidential number. They presumed that their competitor would do the same thing. To keep it all proper, MacGregor and Kukula had to write a computer program, which in those days was punched into cards. They then handed the deck of punchcards across a counter outside the closed computer room. Two days later they would get word back that they had a misplaced comma and their program didn't run. I began to get increasingly agonized phone calls from Ian, who was eventually screaming in frustration that he and Jim were slowly getting nowhere. Gradually, they learned the system and things improved from absolutely execrable to merely bad. Along the way, Jim Kukula located a number of errors in the official data. For instance, in some years the production of all the individual mines did not add up to the claimed total U.S. production. Ian and Jim heard office rumors that the DOE had begun using Jim's numbers as the "official" record. (A few years later, it was reported that the DOE accounting for bomb-grade plutonium and uranium was entered in a ledger book, in pencil, with lots of erasures. We were not the least bit surprised.)

In August, Ian MacGregor went on a family vacation and I replaced him in Grand Junction. At first I didn't fare any better than he had, then a miracle happened. Most of the federal bureaucrats were about to leave on vacation in the last half of August. Before they left, one of them told us that the raw original data were on "tape number 7168, but don't say I told you." Their closed computer room had both a day and a night staff; Jim and I switched to working nights. Every afternoon, I would visit the local bakery and load up on pastries and sodas for midnight snacks to share with the night computer staff. We were behind the counter! In two weeks, Kukula and I extracted enough data to generate a five hundred-page atlas of U.S. uranium production.[11]

It wasn't enough simply to know the amount of uranium in each ore grade. We also needed to know whether the distribution looked the same in different geographic areas, in deposits with different geology, in mines of different depths, and so on. The results were very gratifying:

- The slope of the curves for uranium content versus ore grade was essentially the same for different types of deposits.
- Each time the ore grade dropped to half as rich, there was six times as much uranium in the next slice. An equivalent statement is that dropping the ore grade by a factor of ten uncovered three hundred times as much uranium. One useful fact became evident: The curve was so steep that the breeder reactor was not necessary; there was plenty of uranium to draw upon.
- A single-topped bell-shaped curve described the uranium concentrations in the mines, uranium in noncommercial deposits too lean to mine, and the uranium distribution in common rocks. The proof was in: In this little skirmish Princeton won over Yale.

For a while after our uranium results were published, Brian Skinner was quiet about metal abundance curves with two peaks. However, he's still convinced that other metals may show double-topped peaks. He well may be correct for some metals. I think chromium is likely an example of a metal with almost nothing in between the high-grade deposits and common rocks. This is only a minor battle in a college rivalry that has raged for 250 years. Don't worry, Brian Skinner and I are friends, but at my house, "Yale" is a four-letter word.

Some critics said that uranium abundances looked like one big peak only because there were ten or twenty types of geologically different uranium deposits, each with its own peak. My reply to that is, "Yes!" Bell-shaped distributions usually arise because there are a number of different little causes that add up to one big effect.

We wanted to publicize our findings and the resulting argument against the imminent need for breeder reactors. We chose *Scientific American* because of its wide circulation and its reliance on direct author-ship by the scientists. However, the editor who worked with us patiently explained that the Three Mile Island meltdown had killed all immedi-ate interest in uranium. We had to wait until January 1980 for the arti-cle to be published.

We pause here for a where-are-they-now:

- If you can't beat 'em, join 'em. Ian MacGregor became a fed-eral bureaucrat, distributing research grants for the National Science Foundation. He has recently become the executive di-rector of the National Association of Geoscience Teachers.
- Jim Kukula is a computer scientist with Synopsys, verifying that complex designs for computer chips will do what they are supposed to do.
- Our editor at *Scientific American* moved on to other jobs. He is the editor who developed and nurtured this book.
- I'm retired; my hobby is pushing geology. J. Edgar Hoover has a term in purgatory waiting for me for bribing my way into the Grand Junction computer room with cream puffs.

Radioactive Waste Disposal

Any program, from a geology department to a basketball team, can be disabled by setting the standards impossibly high. In the last twenty years, stringent environmental standards have often been used as a means of defeating a program. Let me confess one of my sins. In the 1970s, the U.S. Air Force wanted to base the MX missile in the Basin and Range province: Nevada and western Utah. Some of us were afraid that the MX was strategically destabilizing. It was a silo buster, a system built to destroy Soviet missiles before they were launched, in short, a first-use weapon. In contrast, the submarine nuclear missiles did not have to be launched immediately in a crisis; they were stabilizing. There were two versions of launch on warning, both of them deeply frightening. One version takes place during the fifteen minutes in be-

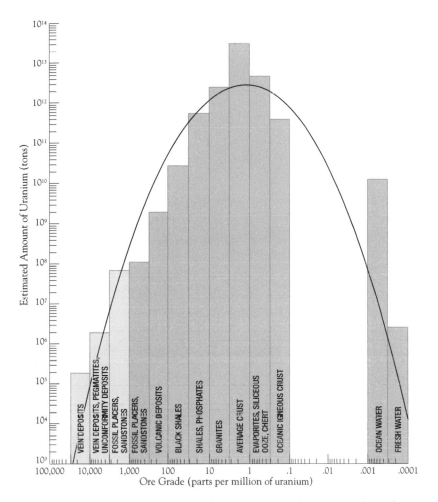

The distribution of uranium in the earth's crust among the major geological reservoirs of the element is plotted here on a log-log chart. The bars representing the various categories of the ore deposits (in descending order of their ore content) define a log-normal global-abundance curve, which in this instance takes the form of a parabola opening downward (black curve). The important point is the slope of the parabola in the lower-left corner. The breeder reactor is not needed because sufficient uranium is available as the ore grades are lowered. (Jeffrey L. Ward)

tween detection of incoming missiles and the time the MX missiles had to be on their way. The other version is political nervousness: Either side could become convinced that its opponent was about to launch a preemptive strike. Added to the nervousness were mistakes, for instance when a flock of geese flew in front of a radar antenna.

My small contribution to the debate was showing that a cancer-causing natural mineral, much worse than asbestos, was found in several of the Nevada basins. It was one of many environmental objections to basing the MX in the Basin and Range province. Eventually, the Reagan administration abandoned the Nevada-Utah MX basing mode, saying that environmentalists could keep them in court forever. I was publicly labeled as "intellectually dishonest" by Antonia Chayes, a former undersecretary of the air force, for using environmental objections because I didn't like the MX as a weapons system. I plead guilty. But I did not falsify any evidence. The cancer-causing mineral really is there. However, I went to the trouble of calling attention to the problem because I thought the MX was a flawed weapon.

Since it takes a crook to catch a crook, let's look at the disposal of radioactive wastes. The first rule of substitute environmentalism—such as my objection to the MX—is setting the standards very high. With radioactive wastes, the standard is simple: They must be stored where they cannot leak out during ten times the longest radioactive half-life in the wastes. Plutonium 239 has a half-life of 24,100 years; ten half-lives amounts to 241,000 years. We've had three ice ages in the last 240,000 years. But just for scale, when do the wastes become less radioactive than the uranium ores that were mined to feed the reactor? About 15,000 years. That's not to say that we can be sloppy about waste disposal; it simply says that the existing barrier has been set as high as possible. Also the scale of the problem is important. If you got 100 percent of your electricity from nuclear power for your entire lifetime, how big would your share of the radioactive waste be? About the size of a doorknob.

A second tactic is substituting a bad answer to the problem, instead of the best solution, and then mounting an attack on the bad approach. Fifty years ago, it was obvious that sedimentary beds of halite (ordinary table salt) and anhydrite (calcium sulfate) were the most effective seals on top of oil and gas reservoirs. Many of the great oilfields of the Mid-

The Yucca Mountain site for radioactive waste disposal is in an unpopulated area adjacent to the western edge of the former Nevada nuclear test site. The derrick on the left is drilling a test hole into the proposed disposal horizon. (U.S. Department of Energy, Science Photo Library)

dle East have layers of anhydrite above the oil and gas. Anhydrite and halite don't leak oil, gas, or even helium. In 1957, a report on the disposal of radioactive wastes discussed using halite or anhydrite beds.[12] The New Mexico site for disposing of military radioactive wastes, called the Waste Isolation Pilot Plant or WIPP, is in a halite (salt) bed. Which brings us to Yucca Mountain.

The original 1972 proposed civilian waste site was in a halite bed in Kansas. "NIMBY!" ("Not in My Back Yard!") shouted Kansans. Eventually the political pressure led to Yucca Mountain, adjacent to the now-inactive Nevada nuclear test site. For years, I stopped my field trip for first-year students on the highway opposite Yucca Mountain and asked, "Whose backyard is this?" You could turn around 360 degrees and the only visible trace of human activity was the highway. Yucca Mountain has faults, literally and figuratively. There is no trace of anhydrite or halite. I'm not saying that Yucca Mountain could not be utilized as a safe site. The NIMBY objections substituted an easy-to-attack Yucca Mountain for the better sites.

Yucca Mountain even produced some comedy. The site is located in Nye County, Nevada. Nye County has a larger area than New Jersey, Massachusetts, Delaware, and Rhode Island combined. Federal money was to go to Nye County as a compensation for accepting the radioactive wastes. The Nevada state government tried to short-circuit that by establishing a new county that included only Yucca Mountain. It was named Bullfrog County, after an early gold mine. Bullfrog County would have no county commissioners, no sheriff, and no budget. All the federal compensation funds would flow through to the state government. Unsuccessfully, I tried to think of some crime that I could commit in Bullfrog County that would go unpunished. The new county never got started; it sounded too much like something Mark Twain made up.

Current Picture

- Electricity generation in France is 77 percent nuclear.
- Cleanup problems from previous U.S. programs run from large (Rocky Flats, Colorado) to gigantic (Hanford, Washington) to I-don't-want-to-think-about-it (Arco, Idaho).
- Nuclear cogeneration of heat, electricity, and electrolytic hydrogen is being considered to assist the oil extraction from Canadian tar sands.
- In June 2003, the U.S. Senate passed a bill providing loan guarantees for the first new U.S. nuclear power stations since 1980.

Waiting for the first new U.S. nuclear power station since 1980 has been like waiting for the Messiah (first or second arrival, your choice). Eventually, it has to happen. The question is whether any of us will live long enough to see it happen.

In my opinion, M. King Hubbert was correct: Expanding the use of nuclear power is an important response to the growing shortage of oil and natural gas. Nuclear power can solve only a portion of the energy problem, but it is a substantial portion. Getting public acceptance of the safety of the reactors and the integrity of the waste disposal will not be easy. Some observers say that a major, and painful, energy shortage has to develop before the public will accept changes. The expansion of nuclear power is just one of the many changes that one can foresee from a vantage point atop Hubbert's peak.

Nine

Hydrogen

Yes, my friends, I believe that water will one day be
employed as fuel, that hydrogen and oxygen which
constitute it, used singly or together, will furnish an
inexhaustible source of heat and light.

—JULES VERNE, *THE MYSTERIOUS ISLAND*, 1874

Hydrogen is not a fuel that comes from the earth. However, this chapter is included because hydrogen has been widely discussed as a cure for the world's energy problems. Using electricity to convert water into hydrogen and oxygen was known for seventy years before Jules Verne wrote about it in 1874. The phrase "hydrogen economy" was coined by General Motors in 1970. In 2002, President Bush announced plans for a hydrogen-powered "Freedom Car," to be available in ten to twenty years. In the same year, Jeremy Rifkin published a book titled *The Hydrogen Economy*.[1] In an enthusiastic moment in 2003, Rifkin made an analogy to "Moore's Law," which states that the number of transistors on an integrated circuit doubles every eighteen months. Rifkin said, "There is an energy revolution happening that is similar to what we saw with the PC revolution. Moore's Law has set in with fuel cells."[2]

There are limits to the magic achievable in the hydrogen economy. We have to ask whether the concept has been oversold. However, there are two pieces of encouraging news:

- There is no absolute scientific barrier to using renewable wind- or solar-generated electricity to produce hydrogen from water.
- Shell has opened a hydrogen filling station in Iceland, with stations in other countries to follow.

Driving around in nonpolluting automobiles, powered by a renewable fuel source, is a consummation devoutly to be wished. Is the hydrogen economy an achievable goal? That question divides into two parts:

- Is hydrogen an effective solution to the problem? Does it take more energy to supply the hydrogen than you get back when you drive the car? As an example, when the SR-71 Blackbird supersonic aircraft was first proposed, hydrogen was suggested as the fuel. Eventually, Kelly Johnson, the head of the legendary Lockheed Skunk Works, decided that he could build a better aircraft using conventional hydrocarbon jet fuel.
- Can we make an orderly transition from our present gasoline-powered cars to a hydrogen fleet? Do we have a chicken-egg problem? Companies will manufacture hydrogen-powered cars if there are hydrogen filling stations. Companies will open hydrogen filling stations if there are hydrogen-powered cars. Here, legitimately, governments can solve the chicken-egg dilemma by using temporary incentives, tax rebates, or subsidies. "Temporary" is the operative word. All too often, tax incentives become permanent. Again, an example: Several analysts have computed that producing ethyl alcohol from corn consumes more fossil fuel energy than you get back when you burn the alcohol. Why do we add alcohol to gasoline? If you want to win the election in a farm state, vote for alcohol subsidies.

The big attraction of hydrogen is the promise of mobility. Roughly 75 percent of our existing oil consumption is for transportation: planes, trains, cars, and boats. Transport of all kinds is an endangered species in an oil shortage. There are numerous uses for hydrogen and major envi-

Refueling a city bus with hydrogen from the Shell station in Reykjavík, Iceland. (Shell Photographic Services, Shell International Ltd.)

ronmental rewards. However, the real reason for considering hydrogen as a fuel is to renew our aging love affair with the automobile.

Currently, no companies are selling hydrogen-powered automobiles. For the Shell hydrogen filling station in Iceland, the initial customers are three prototype city buses. Estimates for initial sales of hydrogen cars range from three to twenty years.

If you had an inexpensive way of making hydrogen, you could get big Texas rich right now. You don't have to wait for hydrogen-powered cars. The global market for hydrogen is about a billion dollars per year. About 85 percent of the hydrogen goes into two uses: making nitrogen fertilizer and upgrading oil in petroleum refineries.

Current Hydrogen Production Practices

The ways of making hydrogen come from chemistry, physics, and (maybe) biology.

- The chemical methods are modern variations on the water-gas process, which is two hundred years old (described in Chapter 5). In today's practice, either natural gas (methane, CH_4) or coal (mostly carbon) burns with air (or pure oxygen) along with steam (water, H_2O) to raise the temperature to 1,000° or 2,000°F. The resulting gases, when cooled, contain a substantial amount of hydrogen, which can be separated and used.
- An electrical current can be passed between two metallic electrodes immersed in water. Bubbles of hydrogen rise up from the negative electrode; oxygen gas is liberated at the positive electrode.
- To date, no biological or biochemical systems have been used commercially, but some bacterial assemblages are reported to produce hydrogen.

Most hydrogen is produced by its consumer at the site where it will be used. Only 3 percent of the hydrogen produced in the United States is bought and sold. The cheapest present-day route for producing hydrogen is the first chemical method listed above: reacting natural gas with steam and oxygen. However, during the spring of 2003, natural gas prices remained high after the preceding cold winter. Several American fertilizer factories, which used hydrogen made from natural gas and nitrogen from the air, were suddenly losing money and were forced to close. That component of the fertilizer industry will have to relocate to places where there is still a surplus of natural gas. That means the associated jobs will shift to western Siberia, in case you are interested in moving.

Water-Gas Variations

Methane, the dominant molecule in natural gas, contains four hydrogen atoms and one carbon. Because natural gas is so hydrogen-rich, oxygen is not added; the reaction involves only methane and steam.

$$CH_4 \quad + \quad 2\,H_2O \quad = \quad CO_2 \quad + \quad 4\,H_2$$

methane   steam   carbon dioxide   hydrogen

The methane and steam are preheated to about 1,500°F and then re-acted in the presence of a nickel catalyst. This reaction also produces substantial amounts of carbon monoxide, which is toxic enough to cause accidental fatalities. A rather complicated set of additional reactors is used, both to convert the carbon monoxide to carbon dioxide and to purify the hydrogen. Heat exchangers use hot gas exiting from the reactors to preheat gas at other places in the system. Despite the complexity of the processing plant, natural gas historically has been the cheapest source of hydrogen. In present practice, the carbon dioxide is simply sent up the smokestack. But, to the extent that hydrogen production could be set up using the associated natural gas from oil production, the waste carbon dioxide would be valuable for boosting oil recovery.

Where natural gas is not available, coal is the second choice. Coal was the original fuel for water-gas production two hundred years ago. The modern hydrogen-from-coal process was developed by Texaco, before it became part of Chevron/Texaco (and Shell has a similar process). As mentioned in Chapter 5, Texaco has sold eight plants to China for fertilizer production. Here's the clever twist in the Texaco design: The object of producing hydrogen is to react it with atmospheric nitrogen to produce ammonia. Our atmosphere contains 78 percent nitrogen and 21 percent oxygen. Since that nitrogen has to be separated from the atmospheric oxygen, the Texaco engineers feed the by-product pure oxygen, instead of air, into the initial coal-steam reaction. Higher temperatures result and atmospheric nitrogen does not get a costly ride through the hydrogen-producing portion of the system. The resulting ammonia can be added directly to irrigation water, converted to crystalline ammonium nitrate, or converted to solid urea; all three products are effective fertilizers. A historical note: Urea was the very first "organic" chemical to be synthesized from inorganic reagents. Before 1828, organic compounds were thought to be produced solely by organisms.

Electrolysis

Producing hydrogen and oxygen from water would seem to be the simplest thing in the world. The process has been known for two hundred years and has been employed commercially for seventy-five of those years. An interesting bit of continuity: Hydrogen for Shell's new filling station in Iceland is generated by electrolytic cells manufactured by Norsk Hydro. In 1943 and 1944, an electrolytic plant in Norway operated by Norsk Hydro was Germany's only source of heavy water (deuterium) for building a nuclear reactor. Two commando raids, a bombing raid, and sabotage kept heavy water from leaving Norway.

There's good news and bad news about producing hydrogen by electrolyzing water. The distinction is between the electrical current (amperes) and the electrical voltage (volts). The good news is that the electrical *current* passing through the electrolytic cells makes hydrogen with better than 98 percent efficiency. The bad news is that the electrical *voltage* required is 20 to 35 percent higher than the theoretical ideal voltage. If the hydrogen is then used in a hydrogen-air fuel cell, much the same thing happens. The electrical current comes out with near 100 percent efficiency, but there is a voltage penalty; you get less than the ideal voltage. In practical terms, the most efficient commercial cells for producing hydrogen require 1.75 to 2.0 volts and the fuel cell returns about 0.7 volt. The bottom line: The hydrogen economy has a built-in surcharge. You get back only about 40 percent of what you put in. Las Vegas gives you better percentage returns.

Could further research improve the voltage efficiency? Not likely. On the production side, a survey of commercial electrolysis systems shows that companies in Norway, Switzerland, Canada, Italy, and Germany have converged on essentially the same chemistry, operating conditions, and efficiency. Possibly because of its long experience, Norsk Hydro is slightly ahead on efficiency. Electrolytic hydrogen production is a mature industry. A 30 percent improvement in efficiency is theoretically possible, but engineering designs stagnated about fifty years ago. There are two ways to react to this. Either this is a golden opportunity for a research breakthrough or be warned that engineers in five different countries converged on the same design.

Small electrolytic hydrogen-generating plants simply buy their electricity from the local power grid. A major plant might develop with its own independent electrical source. In an ideal world, one without economists, an integrated electricity-and-hydrogen facility would be immune to fluctuations in the price of electricity. However, at times of high electricity prices, the integrated plant would be utilizing its electricity at less than market price to subsidize its hydrogen production. During the California energy crisis of 2000–01, some aluminum manufacturers elected to shut down production, giving their workers full pay to stay home, and sell their electricity. While the crisis lasted, they made more profit selling electricity than they would have made selling aluminum. Having your own supply is a short-term insurance against price fluctuations. In the long run, however, you don't waste your potential electrical profit by subsidizing cheap aluminum, or cheap hydrogen.

Since hydrogen can be stored (either as a pressurized gas or as a cold liquid), solar and wind power are particularly attractive for generating hydrogen. But for all the obvious reasons, solar and wind energy are intermittent and not fully predictable. At the time of this writing, wind generators are well engineered, and electrical wind-generating capacity worldwide is expanding at 25 percent per year.

Solar electrical generation is not yet competitive for large-scale facilities. While writing this book, I wanted to build a 60-watt solar-powered demo to produce hydrogen and oxygen. To my surprise, solar cells to produce 60 watts cost $1,000.

Exotic Hydrogen Sources

Over the years, a wide range of methods have been explored for hydrogen production. For instance, breaking down hydrogen-containing chemical compounds (other than natural gas and coal) at high temperatures has been explored, but the processes never made it into commercial practice. On the drawing board are designs for nuclear fission reactors that run at 1,000°F and produce both electricity and hydrogen.

Biological systems in living cells and biochemical systems that mimic biological processes are both interesting. The ideal biological system would capture sunlight, do its photochemical magic, and deliver hydrogen. I have always been a bit dubious. It seemed to me that making methane would be a much smarter move for the bacteria. The long-standing hydrogen champs among biological systems are several species of purple bacteria, whose color reflects a molecule other than green chlorophyll for absorbing sunlight. Passengers approaching the San Francisco airport see many square miles of ponds at the margin of San Francisco Bay, some with a lovely pink lingerie color. The ponds contain evaporating seawater as a means of recovering salt. Their color comes from purple bacteria. Research continues, mostly in Germany, on generating hydrogen by letting purple bacteria work on organic-rich wastewater.

Hydrogen Storage and Transport

As noted earlier, storing hydrogen is also a way of utilizing intermittent solar and wind energy. At the moment, there are three dominant ways of storing hydrogen: as hydrogen gas in high-pressure tanks, as a cold liquid, or as a slush of cold liquid and solid hydrogen. In addition, there is the possibility of storing hydrogen by absorbing it into a liquid or solid host. Even though the first two storage methods are currently in use, they require some extreme conditions. The tank pressure is typically five hundred to one thousand pounds per square inch. If you opt for liquid hydrogen, bear in mind that by "cold" I mean a temperature just above absolute zero, 487°F. The safety issues will be discussed later; let's discuss the engineering.

Storing hydrogen as a high-pressure gas has the advantage of being stable. You can leave the car parked in the garage for a week and the hydrogen is still there. A lower-pressure tank would work, but the tank would fill all the trunk and passenger space in the car. Natural gas–powered cars and trucks have the same problem. Natural gas filling stations have operated for more than twenty years in Italy and in New Zealand, and they have developed an interesting tactic for delivering high-pressure gas. The power required to compress the gas to extreme pressure is fairly large, which costs money. When you pull up to a natural gas filling station with your gas bottles almost empty, it is wasteful to start by taking gas from a tank that has already been compressed to the top pressure. As a consequence, the filling station typically has three compressors and three holding tanks at low, medium, and high pressure. When an almost empty natural gas car pulls up, they start by putting in low-pressure gas, then add medium-pressure gas, and top it off with gas from their highest-pressure tank. Presumably, hydrogen filling stations would operate the same way.

As mentioned, liquid hydrogen is cold, really cold. Of course, the fuel tank in the vehicle is insulated, but no insulation is perfect. Heat is slowly conducted into the tank and hydrogen gas boils off the liquid hydrogen. If the fuel tank were sealed up and you left the car in the garage for a week, when you returned you would find your car and the garage splattered all over the neighborhood. A safe leakage path for the hydrogen, or an on-board refrigeration facility, has to be provided.

With present technology, the break-even point for powering a vehicle with high-pressure hydrogen versus liquid hydrogen occurs at a fuel consumption equivalent to ten gallons of gasoline per day. More than ten gallons per day, you are better off using liquid hydrogen. I was surprised that the crossover amount was so small. It is interesting that the Shell hydrogen station in Iceland is selling high-pressure hydrogen gas; the proposed Shell station in Tokyo will sell cold liquid hydrogen.

There has been a long-standing hope that some kind of absorbent would hold and release hydrogen without requiring either high pressure or low temperature. Some metals do absorb hydrogen. (The 1989 flap involving cold fusion at the University of Utah began with a study of hydrogen absorption into palladium metal.) So far, none of the many proposed absorption schemes has produced a commercial success.

Environmental Issues

A hydrogen economy based on renewable, nonpolluting hydrogen is an enormously attractive environmental goal. Of course, it is important to remember that the hydrogen sources, as well as its consumption, need to be nonpolluting. For instance, an article about hydrogen in the Harvard alumni magazine in early 2004 included a diagram showing natural gas going into a facility and hydrogen and electricity coming out.[3] Natural gas or methane, CH_4, comes in, but nowhere does the carbon emerge. Most existing plants simply send the carbon up the smokestack as carbon dioxide, adding to the atmospheric buildup of carbon dioxide.

Because natural gas demand is starting to exceed the supply in North America, reacting methane with steam is not an option. If a method for producing natural gas from methane hydrates ever became economic, then producing hydrogen from methane might have a future. However, as explained in Chapter 4, in the twenty years after gas hydrate deposits were discovered, no attractive production methods have surfaced.

Hydrogen production from coal, on the other hand, is already a success in China. It would also be an environmental success if the byproduct carbon dioxide were either stored underground permanently or, better yet, used in the recovery of oil. More troubling is the fact that some coal beds contain trace amounts of toxic materials like mercury or arsenic; the hydrogen-from-coal process would have to avoid releasing them to the surface environment. If these problems were surmounted, there is a bright side. The United States and the world in general have coal supplies adequate for a few hundred years.

Electrolyzing water to produce hydrogen is currently (sorry, unintentional pun) about three times as expensive as hydrogen from coal or methane. As explained earlier, serious engineering of electrolytic cells goes back seventy-five years, so don't expect to hear someone in the research lab suddenly shouting "Aha!" We will probably have to accept the present-day efficiency of electrolytic hydrogen production.

Evaluating the environmental impact does not begin with the purchase of power from the electrical-grid. If electrolytic hydrogen is to be an environmental success, expanding the electrical-generating system necessary to produce it has to be an environmental success. Shell's original hydrogen filling station in Iceland is powered by hydroelectric and geothermal electricity. That's because *all* the electric power in Iceland is hydroelectric and geothermal. Solar, wind, and nuclear power can also be nonpolluting. What we cannot do is get the electricity from an existing dirty coal-fired electrical power plant and claim that the environmental bookkeeping begins only after we buy the electricity.

Safety

Sometimes I worry that we overlearn some lessons from history. Munich (1938) and Vietnam (1964–73) are engraved in our collective memory as mistakes never to be repeated. The 1937 burning of the Hindenburg airship taught us, we thought, that hydrogen gas is necessarily disastrous. (The swastikas on the tail fins didn't help.) Two analyses suggest that we learned the wrong lesson from the Hindenburg disaster. The flames in the famous photographs probably show the highly inflammable outer skin burning. There is a curious fact: A hydrogen-air flame is almost invisible. I learned this oddity at the Cordero mercury mine in Nevada. One of the safety procedures at the mine was frequent analyses of the employees' urine to detect low levels of mercury. The mine manager found that the flame analyzer for mercury was even more sensitive if he substituted tanks of hydrogen for the usual natural gas. The only disadvantage of burning hydrogen was a few minor finger burns to the lab technicians. Even after they were told, they instinctively assumed that if they couldn't see the flame the burner was turned off.

A second insight into the Hindenburg lesson comes from a 1954 book by Nevil Shute titled *Slide Rule*.[4] Nevil Shute is remembered for a long list of novels, especially for his apocalyptic *On the Beach*. However, he was originally an aircraft designer and in his one nonfiction book he led a design team that was building a British lighter-than-air, rigid-frame airship. One by one, most of the other rigid airships crashed. Shute flew his to Canada and back. It never crashed. (It was eventually cut up and the frame sold for aluminum scrap.) If you had read *Slide Rule* and had never heard of the Hindenburg, you would wonder why we stopped building rigid airships.

The Apollo missions to the moon carried insulated tanks of liquid oxygen and liquid hydrogen. Fuel cells generated electricity; water produced in the fuel cells was used as drinking water: clever arrangement. However, the *Apollo 13* accident occurred because an electric heater in a liquid oxygen tank got stuck in the on position and the tank explosion blew out one whole side of the spacecraft. It could as easily have been a hydrogen tank with the stuck heater. If so, we might have learned another dubious lesson.

Storing hydrogen in high-pressure tanks on a vehicle is rather similar to storing methane. The problem was discussed in Chapter 4. Little experience exists on the hazards of liquid hydrogen in small vehicles. Hydrogen gas itself is not toxic; as long as a reasonable amount of oxygen is present you can breathe it safely. However, I would suggest leaving your cigarette lighter at home. A wide range of hydrogen concentrations will burn, 4 to 75 percent by volume of hydrogen, the remainder being air. A flame moves through the hydrogen-air mixture at about ten feet per second. There is a second, and more dangerous, combustion mode. In mixtures between 18 and 60 percent of hydrogen in air, an explosion can result: a shock wave that moves at six thousand feet per second. You don't want an exposed flame in a garage where hydrogen has been leaking. On the other hand, think of it as a way of reducing the number of cigarette smokers.

Fuel Cells

Electric decomposition of water into hydrogen and oxygen was discovered in 1800. In 1838, the inverse reaction was demonstrated: Hydrogen and oxygen were recombined to produce electricity. This was not done by burning the hydrogen with the oxygen and using the heat to run an engine. The 1838 device was not limited by the Carnot rule for a heat engine. It became known as a fuel cell. Development continued. The first major success was the Apollo fuel cells.

My first guess was that the insides of a fuel cell would look a whole lot like the guts of an electrolysis cell. Bad guess. The electrolysis cell contains water (usually with dissolved potassium hydroxide to make it conduct electricity) and metal plates hooked up to the electrical source. Bubbles of oxygen grow on the positively charged plates and hydrogen bubbles form on the negative plates. All you have to do is arrange some plumbing to collect (separately) the hydrogen and oxygen bubbles when they rise to the surface. In a fuel cell, electricity is produced where the gas, the water, and the metal plate come together. The insides of a fuel cell are not simple. There are arrangements that work, but when you buy a fuel cell the clever designer and the diligent manufacturer expect to get paid for their efforts. You can buy demonstration size fuel cells and fuel cells that students can take apart.[5]

In both the electrolysis cell and the fuel cell, the composition of the metal surface makes a difference. Platinum is the best of all metals for pushing electrons from the metal surface into, or out of, the water. Platinum sells, as of March 2004, for $10,000 per pound. Electrolysis cells consequently use less expensive steel negative plates and nickel (or nickel-coated) positive plates. Fuel cells need all the help they can get, however, so some use a very thin coat of platinum on top of a cheaper metal underneath.

Most of the world's platinum is produced in South Africa and Russia. If platinum turns out to be an essential component of automotive fuel cells, there will be a squeeze. That's the bad news. The good news is that platinum can be recycled. Today, one of the largest uses of platinum is in catalysts for petroleum refining. Most of the platinum is eventually recovered. I learned about platinum recycling when I was asked to be an expert witness in a lawsuit between a catalyst company and a platinum refinery. I did what most senior faculty members would do: I recruited a graduate student who was in better command of the necessary skills. It was fascinating: "The Case of the Missing Platinum." We gave our depositions and our side obtained a very favorable settlement of the case. The grad student and I wanted to write it up. We were told that the terms of the settlement obligated both sides to keep the evidence confidential. What I can tell you is that most platinum is recycled.

If you are driving a fuel-cell car, cruising along with a light foot on the accelerator, each fuel cell in the stack delivers about 0.7 volt. If you need to pass a truck and you put pedal to metal, the voltage drops to 0.5 volt but the electrical current triples so that the power (current times voltage) doubles. You get around the truck, but at some cost in efficiency. If you try to get even more power—to try to pass a Porsche—the fuel cell cannot move hydrogen and oxygen to the active surfaces fast enough.

The simplest fuel cells use water with dissolved potassium hydroxide, same as the electrolysis cells. However, in search of more efficient

Ford's affiliate, German Ford AG, had this demonstration vehicle operating in February 2001. The hydrogen fuel cells are in the rectangular boxes under the hood. (Jockel Finck, Associated Press)

and more compact fuel cells, half a dozen other types have been engineered. Some run at internal temperatures of several hundred degrees F, which means you won't be able to jump in and start the car right away on a cold morning. Also, there is a search for fuel cells that run on something easier to package than hydrogen. Methanol (methyl alcohol) is one candidate. One potential market is a miniature fuel cell that would run a laptop computer for the entire length of a New York–Tokyo flight.

Of the newer types, the proton exchange membrane, also known as a polymer electrolyte membrane, cell seems closest to going into large-scale production. The fuel cell is divided into two subcells, one fed with hydrogen and one with oxygen (or the oxygen in air). Between the hydrogen and the oxygen subcells is a plastic sheet that will transport protons (hydrogen ions) but not electrons. On the hydrogen side, hydrogen gas breaks down into protons and electrons. Protons move through the plastic membrane and across the cell directly to the oxygen side. Electrons have to detour through a wire connecting the metal plates in each subcell. The electron flow through the wire is the electric power that we use. A Canadian company, Ballard Power Systems, has arrangements to sell proton exchange membrane power cells to several automobile and bus manufacturers.

One proposal calls for replacing a substantial part of the U.S. automobile fleet with fuel-cell cars in ten to twenty years. My concern is with the shorter-term future. World oil supply is already having trouble keeping up with demand. We need a strong plan for coping with the years before fuel-cell cars are a reality.

Lawyers frequently have a way of spoiling the fun; the hydrogen economy is no exception.[6] Currently liability law would probably treat hydrogen as having a "dangerous defect" because of "inability to eliminate the risk by exercise of reasonable care." For nuclear power, the federal government agreed to assume part of the legal liability. Unless the law for hydrogen is revised, better include liability insurance as an additional cost for the hydrogen economy.

One thing certainly will not happen. In his book *The Hydrogen Economy*, Jeremy Rifkin envisions a parking lot full of hydrogen-powered automobiles making money in their off hours by selling electric power to the grid. It is not a result of thermodynamic theory; it is a matter of practical engineering. You paid somebody for 1.7 volts to get the hydrogen and you are going to make money by selling it back at 0.7 volt?

Ten

The Big Picture

An occupational disease that afflicts geologists is thinking that a million years is a short time. The earth formed 4,500 million years ago, so it's hard to be bothered with stuff that lasts less than a million years. Also, there is the illusion (probably wrong) that most of geologic time is wasted. Here are four examples:

- Single-celled organisms occupied the first 80 percent of geologic time. Multicelled plants and animals appeared only during the last 20 percent.
- For 90 percent of the time since multicelled organisms developed, animals other than mammals were dominant.
- 99.8 percent of the age of mammals occurred before anatomically modern humans appeared.[1]
- 90 percent of the time that anatomically modern humans have existed elapsed before agriculture, domesticated animals, and written history were developed.

During the Cenozoic period, the age of mammals, each mammal species lasts for about a million years. That doesn't necessarily mean that on the millionth anniversary, it's "Bang, you're dead" and there are no survivors. Often a species evolves into another species or else splits into two species. Rather than a prediction for extinction, the million-year rule is just a geological hint that *Homo sapiens* isn't likely to go on forever. We might have 900,000 years left. Let's use it wisely. On the bright side, we are different in kind from what went before. Charles Darwin's brain figured out where the brain came from. Because we understand our origins, we have a new opportunity—many would say a new responsibility—to rise above the messy evolutionary process that got us here.

The most recent million years of earth history, called the Pleistocene epoch, have seen multiple advances and retreats of huge continental glaciers, especially around the North Atlantic. Major glaciations are not typical of most of geologic time. (As explained in Chapter 5, the previous set of glacial advances and retreats happened during late Pennsylvanian and early Permian time, 280 to 300 million years ago.) The later stages of human evolution occurred during intermittent glacial periods; we are "children of the Ice Age." The most recent Pleistocene glacial advance extended from about 80,000 years ago to 10,000 years ago. Anatomically modern humans spent most of their 100,000-year history coping with the Ice Age. Archaeological excavations in France show human occupation only forty miles south of the Ice Age glaciers.[2]

Although it is unsettling, sometimes we have to unlearn lessons we were taught in school. When I was an undergraduate, I was taught that Darwinian evolution proceeded fastest when a population was under environmental stress. The opposite is probably true. Any nonlethal mutation can thrive when stress is low, food is plentiful, and the weather is consistent. Diversity survives: Let the good times roll.

The key step that turned a hominid into a modern human is not at all clear. One by one, the characteristics that we thought were exclusively human have turned up in animals and especially in chimpanzees. Jane Goodall showed that chimps deliberately make and use tools. Monkeys use about a twenty-word vocabulary. (An infant monkey saw a leaf falling from a tree and gave the "eagle" warning call.) A wide range of animals learn behavior from watching their parents, which fits one definition of "culture." Chimps and gorillas have been taught sign language by human trainers. However, what none of them seem to have learned is grammar and syntax: rules for the organization of words into meaningful phrases and sentences. As an example, when Washoe, the chimp raised using sign language, first saw a duck on a pond, she signed "water bird." Her human friends thought Washoe meant "water" as an adjective. What Washoe might have said was two nouns: "water," "bird." (I made a similar mistake. A sign on the New Jersey Turnpike says "Elizabeth Goethals Bridge." I knew that George Goethals was a famous engineer; I thought he named the bridge after his wife. Nothing so romantic. The turnoff is to the city of Elizabeth and also to the bridge.) Grammar, the subject almost all of us hated as students, might be the defining human condition. Although it remains to be seen whether humans are different in degree or different in kind, we most certainly are different. The geologically significant fact is our recent arrival. The first agriculture and domestication of animals happened after 99.9998 percent of geologic time had already gone by.

While going around to colleges giving talks about the arrival of the world oil crisis, I have noticed that almost no students are studying real geology: earth history and earth resources. They are taking courses in environmental geology, environmental history, environmental economics, and environmental anything else you want to name. Understanding environmental interactions involves some of the most difficult issues in engineering, chemistry, physics, and computer science. Regrettably, many environmental students aren't interested in the hard stuff; they want to influence policy. This emphasis occurs in a wide range of colleges. As an example, at Princeton we have an endowment to support "a Freshman Seminar in environmental studies, exploring environmental issues primarily through literary, philosophical, ethical, spiritual, or other humanistic perspectives."[3] We've elevated scientific ignorance to an art form.

Global warming is an interesting example. I personally am convinced that global warming is real and we need to reduce carbon dioxide additions to the atmosphere. However, there is an interesting unspoken assumption beneath the global warming debate. The year before the Industrial Revolution (roughly 1720) was the year to which we yearn to return, the year before we began large-scale coal burning. Hold on. In 1720, Europe was still coping with the Little Ice Age. The Little

Ice Age extended from about 1350 to 1850. Before 1350, French city-states had enough surplus productivity to build incredibly impressive cathedrals. Please don't drag me back to a 1720 climate without first thinking about it. Actually, I claim that the Late Miocene might have been a whole lot nicer: The average temperature was about the same, but the poles were warmer and the tropics were cooler than today.

There is now an onging battle for our hearts and minds. Some advisers want us to think of ourselves as caretakers of the earth, with ethical obligations to all life-forms. They say that every little centipede has as much right to thrive as a human. An alternative view says that we won in the Darwinian Olympics, this is *our* planet. We don't want to mess it up and make human life miserable. We also don't want to have another big ice age come along. Try writing an environmental im-

Some of the landscapes painted by Pieter Breugel the Elder are evocative of the Little Ice Age, which extended from about 1350 to 1850. This painting, *The Return of the Hunters*, was done in 1565. (Erich Lessing/Art Resource, New York, Kunsthistorisches Museum, Vienna)

pact statement for a continental glacier thousands of feet thick all the way from Hudson's Bay to south of New York City. This isn't just a Hollywood pitch line. Glacial scratches, only fifteen thousand years old, are visible on rocks in Central Park.

The idealized standard for an environmental impact statement is to have no environmental impact at all. First, do no harm. Humans began making major environmental changes thousands of years ago. The largest impact is from agriculture. Erosion rates, averaged over the entire Mississippi River drainage basin, doubled after farming was introduced.[4] Obviously, it is in our self-interest to reduce agricultural erosion; we want that dirt for growing future crops. Contour plowing, cover crops, and low-tillage farming are examples of erosion control. We are not going back to hunter-gatherer subsistence. We are not about to go back to some idealized former world.

Farming is just the largest of many human modifications of the physics, chemistry, and biology of the earth's surface. The changes are irreversible; we cannot get back to where we started. Viewed from atop Hubbert's peak, we want to find a route to a long-term livable world.

There have been many examples of unplanned global modification. Here is a possible future example of a deliberate instance of global engineering. At least seven ice ages, with major continental glaciers, have occurred in the last million years. Suppose that another glacial advance begins. If we were convinced that oncoming continental glaciers would bury Paris and New York, what could we do? Here's what.

Glacial advances and retreats started about a million years ago, pretty much when the Isthmus of Panama came up out of the water. After the isthmus closed, the trade winds continued to evaporate water off the Atlantic and dump it as rain on the Pacific. Elsewhere around the world, the Atlantic basin and the Pacific-Indian basin are separated by high mountain ranges and major deserts. The water vapor transport across Panama is roughly the size of the Mississippi River; Atlantic water gets saltier and Pacific water becomes more dilute. As salty Atlantic water gets to the Arctic, and especially to the Antarctic, it becomes the densest ocean water and sinks to the bottom. Eventually the cold, salty, dense water fills the bottom 90 percent of the world ocean. Only the uppermost 10 percent of the ocean is available for transporting heat from the equator toward the poles. So how could we prevent further glacial advances? You got it. Move seawater across Panama.

We dig two sea-level canals across the isthmus. Panama has twenty-foot tides on the Pacific side, almost no tides on the Atlantic side. We put flap gates on the Pacific ends of the canals to make one canal flow from Pacific to Atlantic when the tide is high and the other canal flow from Atlantic to Pacific at low tide. In a few hundred years, we wipe out the Atlantic-Pacific salinity difference. The tropics are not as hot, the poles are not as cold, it is San Diego everywhere. I'm not saying this to propose that we do it. I want to make the point that we could make deliberate, as opposed to inadvertent, modifications in the earth's major systems.

Many of the present-day environmental efforts are aimed at preserving biological diversity. "Hotspots" of diversity have been identified; a sure way to get a project funded is to put "hotspot" in the title. Endangered species are protected by U.S. law, a protection that can involve considerable expense. What reasons do we give for protecting diversity? I confess that I am easily seduced by the beauty of diversity. To me, snorkeling on a coral reef or enjoying a magical day on the Serengeti Plain is enormously thrilling. The early geneticist Theodosius Dobzhansky wrote, "Becoming acquainted with tropical nature is, before all else, a great esthetic experience." In part, the high diversity in the tropics reflects environmental stability. The greatest diversity of all is in the waters of the East African lakes, not readily visible to tourists. Lake Victoria contains more than three hundred different species of cichlid fish. (On a trip to Africa, I asked after dinner, "What's a cichlid fish?" "You just ate one," was the answer. That didn't help much; tasted like perch to me.)

At high latitudes, short summers alternate with extremely cold winters. In the Arctic, a whole new crop of mosquitoes has to grow up each year. There are hordes of mosquitoes in the Arctic but only three or four species. Many individuals but few species. Very few tourists go to the Arctic to admire the mosquitoes.

There are several reasons given for protecting biological diversity. One reason is a generalized stewardship: The population explosion of our single species ought not to trigger the extinction of thousands of other species. A more selfish argument says that all species deserve protection because we never know which one might someday produce a chemical that cures cancer or makes better golf balls. If that is our motive, then we ought to be protecting bacteria and fungi. They are the great natural chemists.

Unfortunately, one of the reasons for focusing on diversity is sheer intellectual laziness. It is easy to go through the library counting up how many species of fossils or mosquitoes were reported. It is hard work counting individual rabbits, squirrels, or crows. We're not interested in the success stories. When an immigrant population, like starlings, thrives and multiplies, most of us think it as a disaster. (We have been somewhat more tolerant of the Kennedys.) We love Darwinian losers, species that just barely stay ahead of extinction. There is some good news in the endangered-species department: Killing of African rhinos for the supposed aphrodisiac properties of their horns has diminished. The reason: Viagra.

The Distant Future

The primary point of this book is the disruption of our energy supplies during the next five to ten years. But because I am a geologist, I have an incurable curiosity about the longer time scales. Will we be able to enjoy the remaining 900,000-year life span of a typical mammalian species?

In the depths of the Cold War, it was not at all certain that we had a future. Nevil Shute's On the Beach was about the aftermath of a full-scale United States–Soviet Union thermonuclear exchange. In 1967, after Jocelyn Bell Burnell and Anthony Hewish discovered pulsars, the pulsars looked like a three-dimensional galactic navigation system: that big loran system in the sky! For a few days, I was elated because it meant that at least some civilizations didn't bomb themselves out soon after discovering physics. Within a week, it was clear that pulsars were natural, not operated by extraterrestrial intelligence.

Not long after that, an article titled "The Zoo Hypothesis" was published in the astronomy journal Icarus.[5] Above the article was a note from Carl Sagan, who was then the editor of Icarus. Sagan pointed out that the article was not science because there was no possible way

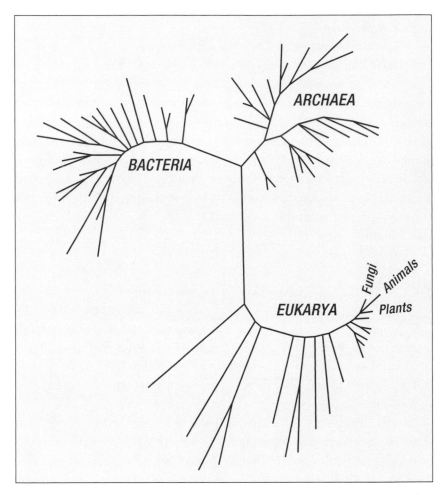

Genetic diversity consists mostly of microorganisms. In this diagram, the lengths
of the lines are the differences between gene sequences in an essential piece
of RNA that exists in all living cells. The lines are grouped to place similar
organisms close together. Bacteria and Archaea have a cell wall enclosing mostly
water with the functional molecules floating free inside. Cells in the Eukarya
domain have internal structures such as nuclei, chloroplasts, and mitochondria.
Visible organisms—plants, animals, and fungi—represent only a small part of the
total diversity. (Jeffrey L. Ward)

to prove the zoo hypothesis wrong. With that teaser, I read the article with interest. The zoo hypothesis suggests that our arm of the galaxy is protected as a nature preserve. Observers from outside would avoid sending in any radio signals or other clues that they are watching. I imagined a bored graduate student somewhere out there coming to work on Monday morning, looking at his computer, and seeing that six civilizations bombed themselves out over the weekend. "That's about average," he thinks and saves the data. Then he starts to read with his feet up on his desk, all five of his feet.

Not long after the end of the Cold War, I stopped having thermonuclear nightmares. Not that the problem is solved; we still have to be exceedingly careful that the nuclear weapons genie stays in the bottle. If we don't bomb ourselves off the planet, what could happen? These are only the rawest of guesses:

- We could detect one, and then several, civilizations around distant stars. We establish contact and become part of a galactic community.
- We never discover advanced civilizations so we decide to colonize some nearby solar systems. Onboard the spacecraft is a major source of human genetic diversity: a refrigerated quart-size container of sperm samples. The crew consists of a few female obstetricians, preferably on the small side.[6]
- We get ever deeper into a symbiosis with computer systems. Silicon is the element right underneath carbon in the chemical periodic table. We're getting married to silicon.

These aren't aims. They aren't recommendations. They probably aren't even right, but it doesn't hurt to think that the present may lead somewhere.

Hundred-Year Timescale

Oil is not the first natural resource to become scarce. The last mine to produce cryolite, used in making aluminum, ran out of ore in 1987. It wasn't a big news item because it was easy to make synthetic cryolite.

An oil shortage is different in kind because we are ill-equipped to generate substitutes quickly. After oil, there will be other materials whose geological supply eventually will exceed the demand. Will there be a natural-resource limitation on Earth's human population? Is there another essential earth material for which there is a limited supply and no substitute? My guess is phosphate. Call it Deffeyes's peak. By discussing phosphate, we can see Hubbert's oil crisis as just one in a series of geological limitations.

Phosphate (PO_4^{+++}) is mined commercially from sedimentary rocks containing the mineral apatite, a form of calcium phosphate. The biggest phosphate producer, with the biggest remaining reserves, is Morocco. By now, you know the message: Major deposits turn out to be concentrated in odd corners of the world. The United States is in second place, with large phosphate mines in Florida and Idaho. Most of the tiny islet of Namur in the Marshall Islands has been dug up to produce phosphate.

We seem to have about a three hundred-year supply of mineable phosphate.[7] Why worry? Won't we find a substitute by then? Phosphate is the backbone of DNA and RNA. The universal energy "currency" within cells is based on the conversion of ATP to ADP, adenosine triphosphate and adenosine diphosphate. Our teeth and bones are made of the mineral apatite.[8] Substitute *that*. The next element below phosphorous in the chemical periodic table is arsenic. Not a promising place to start.

Under any geochemical situation, present or past, the insolubility of apatite limits phosphate dissolved in water to a few parts per million. So how did we get so deeply hooked on phosphate? "We" means all of biology, from viruses to blue whales. Here are two possible answers:

- There is no alternative chemical way to make a viable biological system. If we recover dead microbes from Mars, early on we analyze for phosphate. If we contact a distant civilization on the radio telescopes, we would ask, "Are you made of the atom with fifteen protons?"
- Somewhere there was a microenvironment on Earth with accessible phosphate. I was surprised to find catalog entries for granular column fillings for separating biochemicals. At the

bottom of the list was ground-up apatite. Did our current life-forms originate on the surface of natural apatite grains?

Of course, the answer usually turns out to be "none of the above," but it doesn't cost much to keep thinking about it. I did think about generating the Organization of Phosphate Exporting Countries. Faced with a phosphate shortage, our descendants would do the usual things: recycle phosphate where possible and make up for the unrecycled phosphate by mining low-grade rocks.

The phosphate limitation may not turn out to be the hard limit on human population. Phosphate is used here only as a possible example. The underlying fact is that Hubbert's peak for oil is just one of many resource limitations.

Population Control

Global per capita oil production peaked in 1979.[9] Since 1979, the world has been producing people faster than we have been producing oil. On the long time scale, we have to stabilize our population at a level that the earth can support. Colonizing the galaxy would increase the human population, but only a few pioneers would ride the covered wagons to nearby solar systems. Most people have to stay home. There are two large unanswered questions about the human population: how big and how soon? I'm not certain that convincing answers exist for either of the questions. I would not be surprised if our present-day population has to shrink during the next few hundred years.

Of course, the methods for human population control are enormously controversial. One contraceptive measure seems to be humane and acceptable: If you teach calculus to teenage girls, they go on to have far fewer babies. Calculus is the contraceptive of the future. It doesn't work for boys.

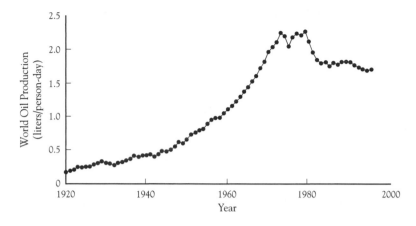

Albert Bartlett showed that plotting the world petroleum production *per person* shows that the peak occurred during the 1970s. Afterward, the increase in production no longer kept up with the rate of population growth. (Jeffrey L. Ward, based on data from *Physics Today*, July 2004)

Future Energy Prices

A universally asked question: What happens to the price of oil? As I've noted, if I had even a partially reliable answer I would be gleefully getting rich playing the oil futures market instead of writing books. The expectation of most economists is that oil will rise to a new equilibrium price. In their models, oil from more expensive sources and reduced consumption will combine to match supply and demand. As pointed out in Chapter 2, it is more likely that oil and natural gas prices will exhibit chaotic price swings.

In one sense, the "inflation-corrected price of oil" is an oxymoron. The price of oil is a major cause of inflation. As we learned in the late 1970s, rising oil prices feed back into higher market prices for almost everything, which in turn generates demands for higher wages. In another sense, the price of a raw material from the earth reflects the skill and diligence of a small population of exploration geologists. To the extent that deposits as rich as those of the nineteenth century are still being discovered, prices remain low. An example: High-grade lead and zinc deposits in Missouri make lead available at 38¢ per pound, zinc at 54¢. In contrast, copper mining has slowly migrated to leaner ores, and the price is up to $1.36 per pound. The number of active exploration geologists in petroleum plus mining in the world is a few thousand, probably fewer than ten thousand. Almost all the students with a natural-science bent today are majoring in environmental studies or ecology. The problem involves more than just the colleges and universities. Most of us learned an enormous amount on the job from our older colleagues, skilled and experienced geologists. When those threads are broken, there is a permanent loss.

Short - Term Problems

After you drive a car off a cliff, it's too late to hit the brakes. In effect, we have gone over the edge of the cliff. In addition to M. King Hubbert's generalized warning, over the last twenty years a dozen different authors predicted that world oil production would peak and start a permanent decline during the 2000–10 decade. Other experts, geologists and economists, dismissed the gloomy forecasts. When the experts disagree, the public usually thinks that no valid knowledge exists. The public response becomes, "What, me worry?"

Hubbert pointed out that Spain was inadvertently devastated by New World gold and silver. During the years it plundered the New World's riches, Spain could sell gold and silver to other countries and buy anything: food, manufactured goods, art objects. By the time the flow of gold and silver ceased, Spain had lost the ability to produce anything and permanently lost its place as a world power. Hubbert's fear was that cheap oil and gas would permanently erode the industrial world's manufacturing and agricultural capability. Economists, in their all-seeing wisdom, point out that the U.S. economy today is much less dependent on energy than it was in 1980. Today, 92 percent of our economy provides services; petroleum is not supposed to matter. I shudder, remembering Hubbert's talking across a cafeteria table about Spain. As long as the world will trade us oil in exchange for Microsoft software and Walt Disney movies, we're in fat city. Hearing that the U.S. service economy is being outsourced to India should be accompanied by Prewett bugling "Taps."

Somewhere, down at the bottom of the economic pyramid, somebody has to be supplying the basic materials. Farming, ranching, commercial fisheries, forestry, mining, and the petroleum industry today are seldom seen as important. Cowboys lost their place as the big movie heroes.

The world oil peak is a world problem. Supertankers can haul oil halfway around the world for two dollars per barrel. Who gets oil from where is largely a matter of transportation convenience. Too many people define the U.S. problem as the amount of oil we import from the Middle East. They are missing the point. Most of Venezuela's and Mexico's oil exports come to the United States. North Africa ships to Europe. Middle Eastern oil moves largely to Japan and to Europe. There is a spot market where shiploads of oil are bought and sold for cash in Rotterdam harbor. Export-import patterns can be rearranged with a few phone calls. As the world oil shortage becomes more severe, who ships what oil where is less relevant.

In the following section, I discuss the potential effects on the U.S. economy. It's not that I'm trying to avoid being a good world citizen. The United States burns 25 percent of the world's oil, so whatever happens in the United States affects the overall world picture.

Communal Action

It looks as if the Hubbert peak is upon us. Whether the maximum year is 2003 or 2005 doesn't matter much. It's real and it's here. Business as usual is not an option. Here I'd better be honest: What we do as a society and what I do as an individual have different constraints. Politically, I'm supposed to believe that my personal moves should be designed to benefit the whole of society. Guess what? I don't live up to that standard. In public, I'll advocate, and vote for, policies that benefit everyone. In private, I confess that I tend to protect my own situation.

Public Policies

With few exceptions, the policies advocated during political campaigns are bland and nonspecific pleas for increased energy efficiency and reduced dependence on imported oil. It is going to be very difficult for the U.S. economy to stand up to an international bidding war for the remaining oil and natural gas. With three-quarters of our own oil already gone, we need something to trade to other countries in exchange for oil and natural gas. Here's the platform you won't hear:

- One of the heroic episodes in American economic history occurred around 1970. Integrated circuit chip manufacturing was moving to Asia. Intel and others won back the market for the most complex and most profitable chips: the central processors. Jet aircraft engines are another instance where American manufacturers provide the highest in high technology. Guessing the identity of the next big technological win is almost impossible, but we need a nurturing environment for innovation. American automobiles are gaining in quality and reliability, a trend to be greatly encouraged.
- A subset of high-technology sales is marketing weapons systems. Selling complex weapons systems to offset oil imports is useful only to a limited degree. If the weapon is so complex that the recipient can't figure out how to use it, maybe it won't do any harm. Don't count on it. The shoulder-launched anti-aircraft rocket was supposed to be complex until someone pointed out that loading and firing a Revolutionary War musket involved a larger number of steps.
- "Conservation" is mostly a euphemism for doing without. If you can't afford it, you will conserve. It's the economists' way of rationing. Unfortunately, the high prices are painful for us ordinary people.
- Regulations requiring improved energy efficiency look attractive on paper. For a government, new efficiency regulations give the appearance of doing something. However, it is not something for nothing. Requiring better automobile highway mileage has to consider the entire system, including petroleum refining.

- We can't drill our way out of the problem. The U.S. oil depletion curve is far past its peak. Even if you give the Arctic National Wildlife Refuge your wildest dreams and open it for drilling, U.S. oil production will continue to decline. Of course, every little bit helps, but ANWR is a very little bit. World oil consumption is roughly 25 billion barrels per year; 5 billion barrels from ANWR would postpone the world decline for two or three months.
- We can no longer conduct contests in which politicians try to outbid one another in promising higher economic growth rates.

I do not present that list to say the situation is hopeless. What the list implies is a profound change in U.S. and world energy usage. My plea is that we go at it with our eyes focused on the long-term goals.

Mining has a place in the larger scheme of things, more than just digging up tar sands. Since 1990, the United States has become the second-largest gold producer in the world. If the state of Nevada were an independent country, it would have the world's third-largest gold production. Critics ask, "What's the use of gold?" The answer as obvious: There are lots of other people overseas who think of gold as money. They are quite willing to trade manufactured goods, or diminishing oil reserves, for those 1,000-ounce gold bullion bars.

Despite numerous troubles, American agriculture has hung in there, remaining globally competitive. High energy prices and the associated high fertilizer prices are a threat to agriculture. Farming methods will have to evolve, but I'm not about to give Iowa back to the buffalo herds. I'd like to see the American farmer be a success story as dramatic as Intel.

Because we ignored the earlier warnings about a global oil peak, an immediate investment in research and development is not an option. There is not enough time. Instead, for the next few years we will have to rely on energy innovations that do not require research and development. Here's my short list of options using existing technology:

- High-efficiency diesel-powered automobiles, getting better than 90 miles per gallon, are now being marketed in Europe. With the new computer controls, diesel engines are not as smoky as before.

- Coal-fired electrical power plants, located where the carbon dioxide can readily be stored underground.
- Wind turbines are now well engineered; the Wyoming wind never seems to stop blowing in Casper.
- Nuclear power plants: How I learned to stop worrying and learned to love the cheap electricity.

We should certainly look for places where two processes can work together. The biggest existing success is cogeneration: burning fossil fuels for the simultaneous generation of heat and electricity. Other possible symbiotic pairs:

- A coal-fired electrical power plant located near an oilfield so that the carbon dioxide can be used to enhance oil recovery.
- A nuclear power plant to supply heat and electrolytic hydrogen for processing heavy-oil sands.
- Agriculture that produces both food and a burnable waste product. An example is bagasse, sugarcane after the sugar-containing juice has been squeezed out. Burning bagasse to produce heat and/or electricity does not add to the atmospheric carbon dioxide burden. The cane plants extracted their carbon from the atmosphere earlier in the same growing season.

Individual Actions

Let's have a private talk around the kitchen table. We won't count on having wealth trickle down from the rich to us ordinary folk. What trickles down from a global energy shortage is going to hurt. Many of the things that we can do to protect our individual welfare are also steps that help globally.

- Transportation: Getting on the wait list to buy a hybrid gasoline-electric car is the most obvious step. Get out from under the long daily automobile commute if you can. (I waited for ten years to find a house two blocks from the university.) Those out-of-season southern hemisphere fruits and vegetables will get costly; learn to use, and treasure, local crops. My favorite first step is making homemade jam, jelly, and preserves during the local produce season.

- Lighting: Get everyone in the habit of turning lights off. An ordinary incandescent lightbulb is a heater that produces small amounts of light as a by-product. Switch to fluorescents, and watch for LED lighting to come on the market.

- Heating and air conditioning: Oil and natural gas prices usually yield about the same heat content per dollar. I have no strong preference. Obviously, natural gas is cleaner, but we're talking survival tactics. Commercially available air-conditioning equipment is all electric. The biggest gain with existing equipment is making optimal use of the outside air. There is a range of temperature and humidity in which most people are comfortable. A winning summer day is when you can keep the windows open during the night, close them after breakfast, and get through the afternoon without turning on the air conditioning. The worst loser is common in New Jersey: The temperature is fine but the relative humidity is 99 percent.

- There are two categories of exhaust fan installations. The small version kicks roof-heated air out of the attic. The large version works alongside the open windows by exhausting hot air from the top of the living space. Both can help. If President Carter didn't convince you (or the previous owner) to upgrade your home's insulation, do it now.

- Anyone with serious energy paranoia can pick up where 1980 left off and build a sod-roofed, solar-heated, ultra-insulated house. Show your sincerity: Build a bike rack instead of a garage.

On the investment side, there are not many obvious tactics. Choose a stock by throwing a dart at the fine-print daily listings, or consult your local economist.

On the downside, aviation is particularly at hazard. Airlines, builders of military and civil aircraft, and jet engine manufacturers are going to face reduced demand if oil prices rise. Experimental hydrogen-

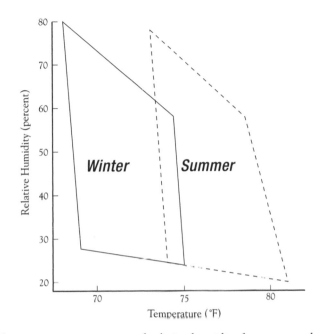

The comfort zone in temperature and relative humidity for most people is the area inside the solid lines for winter clothing and the dashed lines for summer dress. Energy conservation can be encouraged by keeping a copy of this diagram and a relative humidity indicator (hygrometer) near the thermostat. (Jeffrey L. Ward)

powered airplanes were flown as early as the 1920s, but during the next ten years, JP-4, a kerosenelike product distilled from crude oil, will continue to dominate. The engine manufacturers might come out ahead: General Electric and Pratt & Whitney in the United States, Rolls-Royce in Britain. A jet engine's efficiency is determined largely by the compression ratio, the pressure inside the compartment where the fuel is burned divided by the outside air pressure. Early jet engines had compression ratios around two to one. Current jet engines have thirty-to-one compression ratios. A twenty-year-old Boeing 747 looks the same, but the new engines allow it to fly nonstop to the far other side of the world. To the extent that engine designers and manufacturers can squeeze out higher compression ratios, their engines will be retrofitted on existing airframes.

Agriculture is another major concern. (I used to get a serious stomach pain before the evening PBS news. ADM, formerly Archer Daniels Midland, says, "Imagine a world where" ethyl alcohol from corn is a major motor fuel. "Imagine" is the right word. Producing alcohol from corn consumes more fossil fuel energy than you get back by burning the alcohol. The alcohol program works only with heavy federal subsidies.) About 80 percent of an Iowa corn farmer's costs are, directly and indirectly, fossil fuel costs. Producing fertilizer is particularly energy intensive. The nitrogen part of the fertilizer uses natural gas. Rock phosphate undergoes heavy chemical treatment to produce water-soluble phosphate. (Potassium isn't as energy intensive; it is mined in New Mexico and Saskatchewan but has to be hauled to the farms.) Turning the major fertilizer elements into water-soluble forms gives instant gratification to the plants, to the farmer, and to my lawn. However, after a couple of good rains, much of my lawn fertilizer winds up polluting surface and subsurface water, and I'm back buying more fertilizer. A few researchers are trying to reinvent the whole approach to mineral fertilizers.

The Green Revolution of the 1960s and 1970s made famine obsolete. (Pestilence is largely under control, but war and death are still riding.) The Green Revolution is based on improved food crops and heavy use of mineral fertilizer. Converting the world entirely to organic farming requires most people to starve to death. The era of abundant and cheap cow plop ended long before the oil peak. An obvious humanitarian need is to keep fertilizer available and affordable in the Third World.

Agriculture, aluminum, and aviation sound like the front end of the alphabet. What other industries in between aviation and zymurgy are in trouble? Which will thrive? I simply do not know, but taking advantage of the view from atop Hubbert's peak helps answer the question.

Personal Investments

At one difficult point in his presidency, Lyndon Johnson said that he "was like a jackrabbit in a hailstorm, hunkered down and taking it." If your job has already vanished, your retirement funds are wiped out, and you can find nothing better than a minimum-wage job, you have my sympathy. You're hunkered down and taking it. For those somewhat luckier, there are choices to be made.

- The first assumption is that the American political system remains intact. If you don't believe that, you start by buying a .50 BMG bullpup and stockpile plenty of ammo. I'm not that kind of survivalist.
- The second assumption is that the American economy continues to function. I've heard lots of scenarios about the impending meltdown of the economy. Almost all of them end, "And so you better buy gold." Gold is supposed to be the ultimate money, especially if it isn't in your safe-deposit box when the banks all close. I'm not a gold bug, either.

If energy prices are fated to become increasingly unstable and increasingly higher, then direct ownership of energy resources becomes attractive. During the oil crisis of the late 1970s, one of my former students bought a woodlot. He purchased several acres of forest land, intending to harvest enough wood each year to heat his house. I thought it was an excellent idea, until wood-burning stoves began to be restricted because of local air pollution.

Buying a producing oil well is messy, literally and figuratively. I thought that registering and insuring an automobile in New Jersey was bad enough; becoming a legal oil producer in most states is much more complex. Resource trusts, discussed below, are an indirect way of owning oil- and gas-producing wells.

Buying equity in an oil company, by purchasing stock, runs into difficult competition with other buyers. *Oil & Gas Journal* every fall publishes oil and natural gas reserves for U.S. oil companies. Minutes, probably seconds, after the reserves appear in the on-line version of O&GJ, the trading symbols of underpriced companies appear on Wall

Street traders' computer screens. The traders are only trying to stay ahead of the merger and acquisition teams that also are looking for undervalued companies. I can't compete against their resources.

What I did settle on was a royalty trust. There are at least twenty oil and gas royalty trusts traded in the United States. In their original form, royalty trusts bought up from mineral-rights owners their one-eighth to three-sixteenths royalty on oil and gas production. In addition, some operators transferred portions of their share to royalty trusts as a way of securitizing their future income stream. A royalty trust typically has one employee, a lady who takes in checks and writes out checks. The terms of its charter forbid a royalty trust to waste its cash flow by drilling dry holes. Ownership shares in the royalty trusts are bought and sold, some of them on the New York Stock Exchange. Share prices tend to track the market price of oil and gas; the annual cash yield tends to stay around 7 or 8 percent of the share price. There are some small tax advantages for owning royalty trusts in the United States; I am told that Canadians enjoy larger tax advantages.

Conclusion

Business as usual is not in the cards. Muddling through is not strong enough medicine. Whether we like it or not, there will be major re-arrangements in the world economy. It would be more orderly if we were to generate a blueprint for a society constrained by the availability of resources. Then we need a noncatastrophic pathway that takes us from here to that blueprint. Welcome to the post-Hubbert world, the world beyond oil.

Notes

Preface

1. M. King Hubbert, "Energy Resources," in National Research Council, Committee on Resources and Man, *Resources and Man* (San Francisco: W. H. Freeman, 1969), p. 196.
2. Kenneth Deffeyes, *Hubbert's Peak: The Impending World Oil Shortage* (Princeton: Princeton University Press, 2001).

Chapter 1: Why Look Beyond Oil?

1. *Oil & Gas Journal* publishes production and reserve numbers in the last weekly issue of each year. Although *World Oil* is an alternative data source, for continuity with earlier authors I have used the *O&GJ* numbers.
2. M. King Hubbert, "Nuclear Energy and the Fossil Fuels," *American Petroleum Institute Drilling and Production Practice Proceedings* (Spring 1956): 5–75.
3. Craig Hatfield, "Oil Back on the Global Agenda," *Nature* 387 (1997): 121; R. A. Kerr, "The Next Oil Crisis Looms Large—And Perhaps Close," *Science* 281 (1998): 1128–31; Colin Campbell and Jean Laherrere, "The End of Cheap Oil," *Scientific American* 278 (1998): 78–83.

4.   U.S. Geological Survey, World Energy Assessment Team, *U.S. Geological Survey World Petroleum Assessment 2000* (Washington D.C.: USGS 2000), Digital Data Series DDS-60 four CDs.

5.   For example, H. W. Parker, "Demand, Supply Will Determine When World Oil Output Peaks," *Oil & Gas Journal* (February 25, 2002): 40–48.

6.   T. R. Reid, *The Chip* (New York: Random House, 2001), pp. 62–64.

7.   R.W. Apple, Jr., "Four Nations Where Forks Do Knives' Work," *The New York Times*, February 18, 2004, p. F1.

8.   From its founding in 1919 until 1961, Halliburton was a successful oil-service company based in Duncan, Oklahoma. After 1961, Halliburton acquired several construction and contracting companies, including Kellogg, Brown and Root. Most of the negative publicity about Halliburton associated with the Iraq war concerns the activities of the construction and contracting side.

9.   I nominate the University of Texas at Austin, Texas A&M, the University of Oklahoma, the Colorado School of Mines, and leave two spaces for everyone to insert his or her personal favorites.

10.   Matthew Simmons is an enormously valuable independent observer, especially because he does not use Hubbert's methods or data. Simmons heads the largest merchant bank in the oil industry. The graphics for his talks are available at www.simmonsco-intl.com, click on "Matthew R. Simmons."

11.   Alan Greenspan, testimony before the House Committee on Energy and Commerce, June 10, 2003, posted at http://www.federalreserve.gov/boarddocs/testimony/2003/20030610/default.htm.

12.   Stephen Kotkin, "What Is to Be Done?" *FTmagazine* (*Financial Times*, London), March 6, 2004, pp. 16–22.

Chapter 2: Where Oil Came From

1.   Sam Carmalt and Bill St. John, "Giant Oil and Gas Fields" in Michel T. Halbouty, ed., *Future Petroleum Provinces of the World*, Memoir 40 (Tulsa: American Association of Petroleum Geologists, 1986), pp. 11–53.

2.   Edgar Owen, *The Trek of the Oil Finders: A History of Exploration for Petroleum*, Memoir 6 (Tulsa: American Association of Petroleum Geologists, 1975), pp. 11–12.

3.   George Sweet, *The History of Geophysical Prospecting* (Suffolk, UK: Neville Spearman, 1975), pp. 86–88.

4.   Lou Michel and Dan Herbeck, *American Terrorist: Timothy McVeigh and the Oklahoma City Bombing* (New York: HarperCollins, 2001), pp. 215–19.

5.   John Sherman and Murray Roth, "Linux Arrives in Upstream Computing," *The American Oil & Gas Reporter*, November 2001.

6. The Sierra Club is the largest, oldest, and, in my view, best of the conservation organizations. I'm using "Sierra Club" as a metaphor for all environmental organizations. The federal Web site recording offshore drilling accidents is http://www.mms.gov/incidents/blow92.htm#top.

7. T. Irifune et al., "Ultrahard Polycrystalline Diamond from Graphite," *Nature* 421 (2003): 599–600.

8. William Maurer, *Novel Drilling Techniques* (Oxford: Pergamon Press, 1968).

9. Guntis Moritis, "Future of EOR and IOR," *Oil & Gas Journal* 99 (May 14, 2001): 68–73.

10. Abraham Pais, *'Subtle is the Lord . . .': The Science and Life of Albert Einstein* (New York: Oxford University Press, 1982), pp. 216–21.

11. For example, Jean Laherrere, "World Oil Supply—What Goes Up Must Come Down, But When Will It Peak?" *Oil & Gas Journal* 97 (February 1, 2001): 57–64.

12. Leonard Kleinrock, *Queueing Systems* (New York: Wiley, 1976).

13. "Corrected for inflation" probably involves circular reasoning. As we learned (the hard way) during the 1970s, energy costs are an important component of inflation.

Chapter 3: The Hubbert Method

1. Michael Lynch, "Petroleum Resources Pessimism Debunked in Hubbert Model and Hubbert Modeler's Assessment," *Oil & Gas Journal* 101 (July 14, 2003): 38–47.

2. M. King Hubbert, "Techniques of Prediction as Applied to the Production of Oil and Gas," in S. I. Gass, ed., *Oil and Gas Supply Modeling*, Special Publication 631 (Washington, D.C.: National Bureau of Standards 1982), pp. 16–141.

3. F. E. Smith, "Population Growth in *Daphnia magna* and a New Model for Population Growth," *Ecology* 44 (1963): 651–53.

4. U.S. Geological Survey, *World Petroleum Assessment 2000*.

5. For purists: $1.762 = \log(3 + SQR(8))$.

6. The mathematically skilled will recognize that we are doing a numerical integration, and the remainder of this paragraph is determining the integration constant.

7. H. W. Parker, "Demand, Supply Will Determine When World Oil Output Peaks," *Oil & Gas Journal* 100 (November 11, 2002): 40–48.

8. This joke circulated several years before the 2004 downward revision of Shell's reserves. Michael Economides, a petroleum engineer at the University of Houston, claims to have originated it.

9. The Saudi announcement was carried by the Dow Jones Newswire on

March 6, 2003. The key sentence: "Saudi Arabia has told Western government and oil officials that the kingdom's crude oil output has reached its limit at around 9.2 million barrels per day and won't rise further, even with a war looming in Iraq."

10. Colin Campbell, *The Coming Oil Crisis* (Geneva: Multi-Science Publishing Company, 1997), p. 73.

11. Sam Carmalt and Bill St. John, "Giant Oil and Gas Fields" in Michel T. Halbouty, ed., *Future Petroleum Provinces of the World*, Memoir 40 (Tulsa: American Association of Petroleum Geologists, 1986), pp. 11–53.

Chapter 4: Mostly Gas

1. Erle P. Halliburton founded his company based on his method and equipment for cementing the well casing in place.

2. J. F. Kenney et al., "The Genesis of Hydrocarbons and the Origin of Petroleum," *Proceedings of the National Academy of Sciences* 99 (2002): 10979–81.

3. B. Sherwood Lollar, T. D. Westgate, J. A. Ward, G. F. Slater, and G. Lacrampe-Couloume, "Abiogenic Formation of Alkanes in the Earth's Crust as a Minor Source for Global Hydrocarbon Reservoirs," *Nature* 416 (April 4, 2002): 522–24.

4. Rosalind Franklin, "The Study of the Fine Structure of Carbonaceous Solids by Measurement of True and Apparent Densities: Part 1, Coals," *Transactions of the Faraday Society* 45 (1949): 274–86.

5. E. D. Sloan Jr., *Clathrate Hydrates of Natural Gases* (New York: Marcel Dekker, 1990), pp. 464–74.

6. A. M. Tréhu et al., "Three-Dimensional Distribution of Gas Hydrate Beneath Southern Hydrate Ridge: Constraints from ODP Leg 204," *Earth and Planetary Science Letters* 222, issues 3–4 (June 15, 2004): 845–62.

7. Simmons's speeches and papers are available at http://www.simmonsco-intl. com.

8. Martin Gardner, *The Annotated Alice* (New York: Bramhall House, 1960), p. 210.

Chapter 5: Consider Coal

1. Martin Rudwick, *The Great Devonian Controversy* (Chicago: University of Chicago Press, 1975), pp. 78–102.

2. Eric Larsen and Ren Tingjin, "Synthetic Fuel by Indirect Coal Liquefaction," *Energy for Sustainable Development* 7 (2003): 79–102.

3. Alfred Wegener, *The Origin of Continents and Oceans*, 4th rev. ed., trans. John Biram (New York: Dover, [1929] 1966).

4. Alan Smith, A. M. Hurley, and J. C. Briden, *Phanerozoic Paleocontinental World Maps* (Cambridge: Cambridge University Press, 1981), pp. 24–26.
5. R. A. Berner, A. C. Lasaga, and R. M. Garrels, "The Carbonate-Silicate Geochemical Cycle and Its Effect on Atmospheric Carbon Dioxide Over the Past 100 Million Years," *American Journal of Science* 283 (1983): 641–83.
6. John McPhee, *Annals of the Former World* (New York: Farrar, Straus & Giroux, 1998), pp. 29–30.
7. Robert Berner, "Modeling Atmospheric O_2 Over Phanerozoic Time," *Geochimica et Cosmochimica Acta* 65 (2001): 685–94.
8. Andrew Knoll, "Patterns of Extinction in the Fossil Record of Vascular Species," in Matthew Nitecki, ed., *Extinctions* (Chicago: University of Chicago Press, 1984), pp. 23–68.
9. David White, "Some Relations in Origin Between Coal and Petroleum," *Journal of the Washington Academy of Sciences* 5 (1915): 189–212.
10. F. W. Clarke, "The Data of Geochemistry," *U.S. Geological Survey Bulletin* 770 (1924): 95–96.
11. Robert Kleinmann, "The Biogeochemistry of Acid Mine Drainage and a Method to Control Acid Formation," Ph.D. thesis, Princeton University, 1979. Also see U.S. Bureau of Mines Information Circular 8905 (1982), pp. 1–22.

Chapter 6: Tar Sands, Heavy Oil

1. $$\text{API gravity} = \frac{141.5}{\text{density of water @ 60°F}} - 131.5$$
2. As an example, see Keith Skipper, *Petroleum Resources of Canada in the Twenty-first Century*, Memoir 74 (Tulsa: American Association of Petroleum Geologists, 2001), caption to Figure 6, p. 119.
3. Chemical ions, like magnesium with two positive charges, are strongly bonded in the center of the hydrocarbon ring in chlorophyll. Vanadium normally carries a charge of +6, but when combined with two oxygen ions, the vanadyl complex, $VO_2{+}{+}$, bonds nicely as a substitute for the magnesium.

Chapter 7: Oil Shale

1. David Bottjer, *Exceptional Fossil Preservation: A Unique View on the Evolution of Marine Life* (New York: Columbia University Press, 2002).
2. R. A. Gulbrandsen, "Buddingtonite, Ammonium Feldspar in the Phosphoria Formation, Southeastern Idaho," *U.S. Geological Survey Journal of Research* 2 (1974): 693–97.

3.   Lawrence Hardie and Hans Eugster, "The Evolution of Closed-Basin Brines," *Mineralogical Society of America*, Special Paper 3 (1970): 273–90.

4.   Charles Milton and Joseph Fahey, "Green River Mineralogy—A Historical Account," *Wyoming Geological Association Guidebook* (1960): 159–62.

5.   Yildram Dilek and Eldridge Moores, "A Tibetan Model for the Early Tertiary Western United States," *Journal of the Geological Society* (London) 156 (1999): 929–41.

6.   Brad Lemley, "Anything Into Oil," *Discover* 24 (May 2003): pp. 50–57, reports a new, patented process for using hot water in a pressure vessel to recover oil from everything from food wastes such as "turkey guts" to anthracite coal. I'm not a patent attorney, but I would look closely at the published methods for treating oil shale.

Chapter 8: Uranium

1.   Niels Bohr and John Wheeler, "The Mechanism of Nuclear Fission," *Physical Review* 56 (1939): 426–50.

2.   Eugene Wigner, *The Recollections of Eugene P. Wigner* (New York: Plenum, 1992), pp. 6–8.

3.   Richard Rhodes, *Dark Sun: The Making of the Hydrogen Bomb* (New York: Simon & Schuster, 1995), pp. 462–63.

4.   A modern discussion of Carnot's result is given in Gilbert N. Lewis and Merle Randall, *Thermodynamics*, 2d ed. (New York: McGraw-Hill, 1961), pp. 94–98.

5.   George Dyson, *Project Orion: The True Story of the Atomic Spaceship* (New York: Henry Holt, 2002), pp. 42–45.

6.   Leslie R. Groves, *Now It Can Be Told: The Story of the Manhattan Project* (New York: Harper, 1962), pp. 34–37.

7.   Robert Hargraves, "Cross-Bedding and Ripple-Marks in Main-Bird Quartzites in the East Rand Area: A Reconnaissance Study," Information Circular 5 (1961).

8.   Uranium analyses are traditionally reported as U_3O_8, which is a mixed-valence compound. Because uranium is so heavy, the oxygen doesn't make much difference. U_3O_8 is 85 percent uranium by weight.

9.   Donella Meadows, Dennis Meadows, Jorgen Randers, and William Behrens III, *The Limits to Growth* (New York: Universe Books, 1972).

10.   Brian Skinner, "A Second Iron Age Ahead?" *American Scientist* 64 (1976): 258–69.

11.   Kenneth Deffeyes, Ian MacGregor, and James Kukula, *Uranium Distribution in Mined Deposits and in the Earth's Crust*, final report prepared for the Department of Energy, Grand Junction, Colorado, 1978. Project reports like this are called "gray literature" because they can be almost impossi-

ble to locate. Copies of this report are archived in the Princeton University Library.

12. Harry Hess, chairman, *The Disposal of Radioactive Wastes on Land* (Washington, D.C.: National Academies Press, 1957), pp. 12–81. M. King Hubbert was a member of the committee.

Chapter 9: Hydrogen

1. Jeremy Rifkin, *The Hydrogen Economy* (New York: Penguin Putnam, 2002).
2. Quoted in Erik Rhey, "Fuel Cells," *PC Magazine* 22 (July 2003): 92.
3. Craig Lambert, "The Hydrogen-Powered Future," *Harvard* 106 (January 2004): 30–35.
4. Nevil Shute, *Slide Rule: The Autobiography of an Engineer* (London: House of Stratus, 2000).
5. About thirty-five different types of demonstration-scale fuel cells, along with related devices, are sold by the Fuel Cell Store, PO Box 4038, Boulder, CO 80306, http://www.fuelcellstore.com.
6. Russell Moy, "Liability and the Hydrogen Economy," *Science* 301 (2003): 47.

Chapter 10: The Big Picture

1. For the smallest of subdivisions, the concept of genus and species doesn't work well. The phrase "anatomically modern humans" refers to skeletons that are not distinguishable from the bones of living humans. In round numbers, anatomically modern humans emerged about 100,000 years ago.
2. Sheldon Judson, "Geological and Geographical Setting (Les Eyzies)" *American School of Prehistoric Research Bulletin* 30 (1975): 19–26.
3. It is the Henry David Thoreau Freshman Seminar in Environmental Studies, listed at http://www.princeton.edu/pr/pub/fs/03/73.htm.
4. Sheldon Judson, Kenneth Deffeyes, Robert Hargraves, *Physical Geology* (Englewood Cliffs, N.J.: Prentice-Hall, 1976), p. 376.
5. John Ball, "The Zoo Hypothesis," *Icarus* 19 (1973): 347–49.
6. Greetings to Dr. Sarah Hougen Poggi, a friend and former student now on the obstetrics faculty at Georgetown Medical School, who is 5 feet 4 inches tall.
7. Frederick Wells, *The Long-Run Availability of Phosphorous* (Baltimore: Johns Hopkins Press, 1975), pp. 9–19.
8. The chemical formula for apatite is $Ca_5(OH,F,Cl)(PO_4)_3$. Inside the first parentheses, OH (hydroxide), F (fluoride), and Cl (chloride) can sub-

stitute for one another. Apatite with OH is almost insoluble in water. Substituting fluoride makes it even more insoluble. That's why having small amounts of fluoride in drinking water helps to prevent cavities. Our teeth are made of apatite.

9.   Albert Bartlett, "Thoughts on Long-Term Energy Supplies: Scientists and the Silent Lie," *Physics Today* 57 (July 2004): 53–55.

Index